Springer Proceedings in Physics

Volume 440

Indexed by Scopus

The series Springer Proceedings in Physics, founded in 1984, is devoted to timely reports of state-of-the-art developments in physics and related sciences. Typically based on material presented at conferences, workshops and similar scientific meetings, volumes published in this series will constitute a comprehensive up to date source of reference on a field or subfield of relevance in contemporary physics. Proposals must include the following:

- Name, place and date of the scientific meeting
- A link to the committees (local organization, international advisors etc.)
- Scientific description of the meeting
- List of invited/plenary speakers
- An estimate of the planned proceedings book parameters (number of pages/articles, requested number of bulk copies, submission deadline).

Please contact:

For Americas and Europe: Dr. Zachary Evenson; zachary.evenson@springer.com
For Asia, Australia and New Zealand: Dr. Loyola DSilva; loyola.dsilva@springer.com

Peter Smowton · Thomas Krauss
Editors

Proceedings of the 26th European Conference on Integrated Optics

ECIO 2025, June 10–12, Cardiff University

Springer

Editors
Peter Smowton
Cardiff University
Cardiff, UK

Thomas Krauss
University of York
York, UK

ISSN 0930-8989 ISSN 1867-4941 (electronic)
Springer Proceedings in Physics
ISBN 978-3-032-29672-6 ISBN 978-3-032-29673-3 (eBook)
https://doi.org/10.1007/978-3-032-29673-3

This Springer imprint is published by the registered company Springer Nature Switzerland AG
The registered company address is: Gewerbestrasse 11, 6330 Cham, Switzerland

Preface

Cardiff University and the City of Cardiff hosted this 26th Edition of the European Conference on Integrated Optics (ECIO) between June 10 and 12, 2025. The program consisted of an exciting mix of cutting edge science in a venue that demonstrates investment in the future and custodianship of the best of the past. As well as exchanging new ideas and meeting old and new friends, attendees had the opportunity to explore the heritage and the culture of Wales within its Capital City.

ECIO has a long history of providing a forum for presenting and discussing the latest developments in the field. This year, a dominant theme was heterogeneous integration, which has become a major driver for combining active and passive functionality on a chip, alongside the established growth-based techniques; both of these are now being used to great effect. Another clear trend was the continued development of quantum photonic integrated circuits, with novel ideas emerging especially in single photon sources. Various realisations of photonic processors were also being discussed, with both the quantum and the neural network- based architectures addressing the long-held vision of an "optical computer". All of these developments are underpinned by a solid foundation in integrated optics, which is a field that has now well matured but still offers surprises. The technical sessions were augmented by lively discussions around "AI for Photonics" and "ECR networking".

While ECIO focusses on materials and device physics and engineering, South Wales contributes an ecosystem based on Compound Semiconductor materials. The CS cluster has generated more than a billion pounds of public and private sector investment with major new research and manufacturing facilities, with companies such as IQE, KLA, Vishay and Microchip manufacturing in the region. Cardiff University has invested in its Institute for Compound Semiconductors (ICS) with a 1500m^2 cleanroom facility fabricating CS devices and integrated circuits on up to 200mm diameter wafers. Some of the local companies and institutes were represented in the exhibition, which ran alongside lunch and refreshment breaks. We thank the exhibitors and our sponsors for their presence and support, which was greatly appreciated by the attendees.

The meeting provided three days of excellent science and the attendees very much enjoyed the meeting and making new friends!

York, UK Thomas Krauss
Cardiff, UK Peter Smowton

Contents

Contributors

Ksenia Abrashitova SMART Photonics: High Tech Campus, Eindhoven, The Netherlands

Tarni Aggarwal Eindhoven University of Technology, Eindhoven, The Netherlands

Luc Augustin SMART Photonics: High Tech Campus, Eindhoven, The Netherlands;
Photonic Integration Group, Eindhoven University of Technology (TU/E), Eindhoven, The Netherlands

M. W. Balkie Fraunhofer Institute for Microelectronic Circuits and Systems (IMS), Duisburg, Germany

Erwin A. J. M. Bente Photonic Integration Group, Eindhoven University of Technology, Eindhoven, The Netherlands

Erwin J. W. Berenschot MESA+ Institute, University of Twente, Enschede, Netherlands

Sandra Borzek Fraunhofer Institute for Laser Technology ILT, Aachen, Germany

Carsten Brenner Photonics and Terahertz Technology, Ruhr University Bochum, Bochum, Germany

Jacob F. F. Bulmer Quantum Engineering Technology Labs, School of Physics and Department of Electrical, Electronic and Mechanical Engineering, University of Bristol, Bristol, UK

Roman Burkard Department of Electronic Components and Circuits (EBS), University of Duisburg-Essen (UDE), Duisburg, Germany

Cosimo Calò III-V Lab, Palaiseau, France

G. Cannetti POLIFAB, Via Giuseppe Colombo, Milan, Italy

Marcelo Carnica Department of Electronics, Information and Bioengineering, Politecnico di Milano, Milan, Italy

Patrícia Carvalho PICadvanced, S.A., PCI—Creative Science Park, Ílhavo, Portugal

Gabriele Cavicchioli Department of Electronics, Information and Bioengineering, Politecnico di Milano, Milan, Italy

Léa Chaccour PITC, Eindhoven University of Technology (TU/E), Eindhoven, The Netherlands

Monica Crico Department of Electronics, Information and Bioengineering, Politecnico di Milano, Milan, Italy

Alberto G. Curto Photonics Research Group, Department of Information Technology, Ghent University-Imec, Ghent, Belgium

Bijoy Krishna Das Centre for Programmable Photonic Integrated Circuits and Systems, Department of Electrical Engineering, IIT Madras, Chennai, India

Edward C. R. Deacon Quantum Engineering Technology Labs, School of Physics and Department of Electrical, Electronic and Mechanical Engineering, University of Bristol, Bristol, UK

Samuele De Gaetano Department of Electronics, Information and Bioengineering, Politecnico di Milano, Milan, Italy

Folke Dencker Institute of Micro Production Technology, Leibniz University Hannover, Garbsen, Hannover, Germany

Sascha de Wall Institute of Micro Production Technology, Leibniz University Hannover, Garbsen, Hannover, Germany

Alessandro di Tria Department of Electronics, Information and Bioengineering, Politecnico di Milano, Milan, Italy

Marcus Dornetshumer EV Group, Sankt Florian ann Inn, Austria

Paulo Duarte PICadvanced, S.A., PCI—Creative Science Park, Ílhavo, Portugal

Florian Dubois Swiss Center for Electronics and Microtechnology (CSEM), Neuchâtel, Switzerland

Moritz Eggeling AIT Austrian Institute of Technology GmbH, Vienna, Austria

Giorgio Ferrari Department of Physics, Politecnico di Milano, Milan, Italy

Gilles Feugnet Thales Research and Technology, Palaiseau, France

Thomas Geruschke Fraunhofer Institute for Microelectronic Circuits and Systems (IMS), Duisburg, Germany

Pietro Giannoccaro Department of Electronics, Information and Bioengineering, Politecnico di Milano, Milan, Italy

Mariangela Gioannini Dipartimento di Elettronica e Telecomunicazioni, Politecnico di Torino, Torino, Italy

Víctor J. Gómez Nanophotonics Technology Center, Universitat Politècnica de València, Valencia, Spain

Melissa Goodwin MESA+ Institute, University of Twente, Enschede, Netherlands

E. Gubello DEIB, Politecnico di Milano, Milan, Italy

Rainer Hainberger AIT Austrian Institute of Technology GmbH, Vienna, Austria

Ben Hasenson PITC, Eindhoven University of Technology (TU/E), Eindhoven, The Netherlands

Julia Hauser Fraunhofer Institute for Microelectronic Circuits and Systems (IMS), Duisburg, Germany

Martijn J. R. Heck Department of Electrical Engineering, Eindhoven University of Technology, Eindhoven, The Netherlands

Hendrik Heine Institute of Quantum Optics, Leibniz University Hannover, Hannover, Germany

Martin R. Hofmann Photonics and Terahertz Technology, Ruhr University Bochum, Bochum, Germany

George M. Jandu Cardiff University, Cardiff, UK

Yuqing Jiao Eindhoven University of Technology, Eindhoven, The Netherlands

Rachel A. Jones Photonic Integration Group, Eindhoven University of Technology, Eindhoven, The Netherlands

Manuel Jung Fraunhofer Institute for Laser Technology ILT, Aachen, Germany

Chengpeng Kang School of Optoelectronic Engineering and Instrumentation Science, Dalian University of Technology, Dalian, China

Alexander Kassner Institute of Micro Production Technology, Leibniz University Hannover, Garbsen, Hannover, Germany

Julian Konig Photonic Integration Group, Eindhoven University of Technology, Eindhoven, The Netherlands

Jannik Korte Fraunhofer Institute for Microelectronic Circuits and Systems (IMS), Duisburg, Germany

Jana Köller Fraunhofer Institute for Laser Technology ILT, Aachen, Germany

Lisa C. Kreuzer Photonics and Terahertz Technology, Ruhr University Bochum, Bochum, Germany

Anthony Laing Quantum Engineering Technology Labs, School of Physics and Department of Electrical, Electronic and Mechanical Engineering, University of Bristol, Bristol, UK

Sylwester Latkowski PITC, Eindhoven University of Technology (TU/E), Eindhoven, The Netherlands;
Photonic Integration Group, Eindhoven University of Technology (TU/E), Eindhoven, The Netherlands

Xaveer J. M. Leijtens Photonic Integration Group, Eindhoven University of Technology, Eindhoven, The Netherlands

Julian Lemburg Institute of Quantum Optics, Leibniz University Hannover, Hannover, Germany

Jacopo Leo Swiss Center for Electronics and Microtechnology (CSEM), Neuchâtel, Switzerland

Jean-Pierre Locquet Department of Physics and Astronomy, KU Leuven, Leuven, Belgium

Alexandre Maïnos Quantum Engineering Technology Labs, School of Physics and Department of Electrical, Electronic and Mechanical Engineering, University of Bristol, Bristol, UK

Pablo Marin-Palomo Brussels Photonics Team (B-PHOT), Vrije Universiteit Brussel, Brussel, Belgium

Fernando Martín-Romero Nanophotonics Technology Center, Universitat Politècnica de València, Valencia, Spain

A. Martinez DEIB, Politecnico di Milano, Milan, Italy

Maria Mayr EV Group, Sankt Florian ann Inn, Austria

Andrea Melloni Department of Electronics, Information and Bioengineering, Politecnico di Milano, Milan, Italy

Bernat Molero Agudo Department of Electrical Engineering, Eindhoven University of Technology, Eindhoven, The Netherlands

Juan-Francisco Morcillo Nanophotonics Technology Center, Universitat Politècnica de València, Valencia, Spain

Francesco Morichetti Department of Electronics, Information and Bioengineering, Politecnico di Milano, Milan, Italy

Geert Morthier Photonics Research Group, INTEC Department, Ghent University, Ghent, Belgium

Paul Muellner AIT Austrian Institute of Technology GmbH, Vienna, Austria

Jonathan Müller Fraunhofer Institute for Laser Technology ILT, Aachen, Germany

Aleksandar Nesic Fraunhofer Institute for Microelectronic Circuits and Systems (IMS), Duisburg, Germany

Joep Nieuwdorp SMART Photonics: High Tech Campus, Eindhoven, The Netherlands

Marco Novarese Dipartimento di Elettronica e Telecomunicazioni, Politecnico di Torino, Torino, Italy

Pawan Kumar Pandit Centre for Programmable Photonic Integrated Circuits and Systems, Department of Electrical Engineering, IIT Madras, Chennai, India

Jorge Parra Nanophotonics Technology Center, Universitat Politècnica de València, Valencia, Spain

Suvarna Parvathy Centre for Programmable Photonic Integrated Circuits and Systems, Department of Electrical Engineering, IIT Madras, Chennai, India

Pepijn W. H. Pinkse MESA+ Institute, University of Twente, Enschede, Netherlands

Andrzej Polatynski VPIphotonics GmbH, Berlin, Germany;
Warsaw University of Technology, Warsaw, Poland

Yasser Pordeli MESA+ Institute, University of Twente, Enschede, Netherlands

Kristif Prifti VTEC Lasers & Sensors, Eindhoven, The Netherlands

Ernst Maria Rasel Institute of Quantum Optics, Leibniz University Hannover, Hannover, Germany

André Richter VPIphotonics GmbH, Berlin, Germany

Cristina Rimoldi Dipartimento di Elettronica e Telecomunicazioni, Politecnico di Torino, Torino, Italy

Carla Rodrigues PICadvanced, S.A., PCI—Creative Science Park, Ílhavo, Portugal

Francisco Rodrigues PICadvanced, S.A., PCI—Creative Science Park, Ílhavo, Portugal

Jakub Sadilek MESA+ Institute, University of Twente, Enschede, Netherlands

Salvatore Salpietro Dipartimento di Elettronica e Telecomunicazioni, Politecnico di Torino, Torino, Italy

Marco Sampietro Department of Electronics, Information and Bioengineering, Politecnico di Milano, Milan, Italy

Pablo Sanchis Nanophotonics Technology Center, Universitat Politècnica de València, Valencia, Spain

João Santos PICadvanced, S.A., PCI—Creative Science Park, Ílhavo, Portugal; Instituto de Telecomunicações (IT), Universidade de Aveiro, Aveiro, Portugal; Departamento de Electrónica, Telecomunicações e Informática (DETI), Universidade de Aveiro, Aveiro, Portugal

Hamed Sattari Swiss Center for Electronics and Microtechnology (CSEM), Neuchâtel, Switzerland

Anna Lena Schall-Giesecke Fraunhofer Institute for Microelectronic Circuits and Systems (IMS), Duisburg, Germany;
Department of Electronic Components and Circuits (EBS), University of Duisburg-Essen (UDE), Duisburg, Germany;
Center for Nanointegration Duisburg-Essen (CENIDE), Duisburg, Germany

Anton Schillings Fraunhofer Institute for Laser Technology ILT, Aachen, Germany

Waldemar Schmidt Rosenberger Hochfrequenztechnik GmbH & Co. KG, Fridolfing, Germany

Koen Schouteden Department of Physics and Astronomy, KU Leuven, Leuven, Belgium

Niklas Schulz Photonics and Terahertz Technology, Ruhr University Bochum, Bochum, Germany

Jan-Jonas Schumacher AIT Austrian Institute of Technology GmbH, Vienna, Austria

Irwan D. Setija Department of Electrical Engineering, Eindhoven University of Technology, Eindhoven, The Netherlands;
Department of Research, ASML Netherlands B.V., Veldhoven, The Netherlands

Alexandre Shen III-V Lab, Palaiseau, France

Georgios Sinatkas Institute for Integrated Photonics IPH, RWTH Aachen, Aachen, Germany

Peter M. Smowton Cardiff University, Cardiff, UK

Franciscus H. B. Somhorst MESA+ Institute, University of Twente, Enschede, Netherlands

Nils Surkamp Photonics and Terahertz Technology, Ruhr University Bochum, Bochum, Germany

Joana Tatá PICadvanced, S.A., PCI—Creative Science Park, Ílhavo, Portugal

António Teixeira PICadvanced, S.A., PCI—Creative Science Park, Ílhavo, Portugal;

Instituto de Telecomunicações (IT), Universidade de Aveiro, Aveiro, Portugal; Departamento de Electrónica, Telecomunicações e Informática (DETI), Universidade de Aveiro, Aveiro, Portugal

Molly A. Thomas Quantum Engineering Technology Labs, School of Physics and Department of Electrical, Electronic and Mechanical Engineering, University of Bristol, Bristol, UK

Roald M. Tiggelaar MESA+ Institute, University of Twente, Enschede, Netherlands

Hon Ki Tsang Department of Electronic Engineering, The Chinese University of Hong Kong, Shatin, New Territories, Hong Kong SAR, China

Kees van der Zouw MESA+ Institute, University of Twente, Enschede, Netherlands

Inês Venâncio PICadvanced, S.A., PCI—Creative Science Park, Ílhavo, Portugal

Martin Virte Brussels Photonics Team (B-PHOT), Vrije Universiteit Brussel, Brussel, Belgium

Konstantinos Vyrsokinos School of Physics, Aristotle University of Thessaloniki, Thessaloniki, Greece;
Center for Interdisciplinary Research and Innovaton, Aristotle University of Thessaloniki, Thessaloniki, Greece

Zelu Wang Department of Electronic Engineering, The Chinese University of Hong Kong, Shatin, New Territories, Hong Kong SAR, China

Marcus Westhues Fraunhofer Institute for Microelectronic Circuits and Systems (IMS), Duisburg, Germany

Zhenlin Wu School of Optoelectronic Engineering and Instrumentation Science, Dalian University of Technology, Dalian, China

Marc Christopher Wurz Institute of Micro Production Technology, Leibniz University Hannover, Garbsen, Hannover, Germany

Yunjie Yan School of Optoelectronic Engineering and Instrumentation Science, Dalian University of Technology, Dalian, China;
Photonics Research Group, INTEC Department, Ghent University, Ghent, Belgium

Patrick Yard Quantum Engineering Technology Labs, School of Physics and Department of Electrical, Electronic and Mechanical Engineering, University of Bristol, Bristol, UK

Francesco Zanetto Department of Electronics, Information and Bioengineering, Politecnico di Milano, Milan, Italy

Homa Zarebidaki Swiss Center for Electronics and Microtechnology (CSEM), Neuchâtel, Switzerland

Panagiotis Zdoupas School of Physics, Aristotle University of Thessaloniki, Thessaloniki, Greece;
Center for Interdisciplinary Research and Innovaton, Aristotle University of Thessaloniki, Thessaloniki, Greece

L. Zerboni DEIB, Politecnico di Milano, Milan, Italy

Mingshan Zhao School of Optoelectronic Engineering and Instrumentation Science, Dalian University of Technology, Dalian, China

Chapter 1
A Customizable 8” Wafer-Scale Photonic Platform for Post-CMOS Integration

Marcus Westhues, Thomas Geruschke, Julia Hauser, Roman Burkard, Jannik Korte, Aleksandar Nesic, and Anna Lena Schall-Giesecke

Abstract We report on the photonic technological development platform at Fraunhofer IMS, which is based on 200 mm wafers. To allow for post-CMOS compatible processes and a tight-knit interplay between established CMOS circuitry and photonic integrated circuits (PICs), we only employ deposition techniques at low temperatures. We demonstrate the viability of a post-CMOS compatible AlN platform, enabled by low-temperature sputtering. We fabricate and experimentally characterize waveguides and ring-resonators in this platform. In addition, we supplement our selection of processes with chemical-mechanical polishing (CMP), which opens additional possibilities for photonic integration. Furthermore, we present our latest results on advancing our silicon nitride platform portfolio with additional devices—as an example we show a waveguide crossing in the C-band, and a 1×2 MMI in the visible spectrum. Finally, we give an outlook on our future development plans.

1.1 Introduction

Integrated photonics has evolved into a broad field with many different routes of development being pursued—from processes based on established silicon technology [1, 2] to III-V materials like, e.g., indium phosphide [3]. With the necessity arising for co-integration of electronic and photonic devices, e.g., to electronically control

M. Westhues (✉) · T. Geruschke · J. Hauser · J. Korte · A. Nesic · A. L. Schall-Giesecke
Fraunhofer Institute for Microelectronic Circuits and Systems (IMS), Duisburg, Germany
e-mail: marcus.westhues@ims.fraunhofer.de

T. Geruschke
e-mail: thomas.geruschke@ims.fraunhofer.de

R. Burkard · A. L. Schall-Giesecke
Department of Electronic Components and Circuits (EBS), University of Duisburg-Essen (UDE), Duisburg, Germany

A. L. Schall-Giesecke
Center for Nanointegration Duisburg-Essen (CENIDE), Duisburg, Germany

P. Smowton and T. Krauss (eds.), *Proceedings of the 26th European Conference on Integrated Optics*, Springer Proceedings in Physics 440,
https://doi.org/10.1007/978-3-032-29673-3_1

active photonic devices, the compatibility and scalability of such processes becomes of importance. Harnessing the scalability of established processes based on silicon technology can only be achieved via post-CMOS integration, where photonics can be directly deposited and fabricated on top of finished CMOS circuits.

In this paper, we present the latest results achieved in our 200 mm wafer-scale dielectric photonic platform for post-CMOS photonics [4]. We introduce the integration of sputtered, post-CMOS compatible aluminum nitride (AlN). We measure propagation losses of 2 dB/cm at a wavelength of 1620 nm, and a ring resonator with an extinction ratio of 15.4 dB and a Q factor of approx. 35,000 in the C-band. Furthermore, we develop our SiN platform by designing, fabricating and characterizing additional photonic devices in the visible spectrum and in the C-band. Here, we present two exemplary devices: a 1×2 multi-mode interferometer (MMI) for 50/50 splitting, which exhibits an estimated 0.12 dB excess loss at a wavelength of 633 nm, and a waveguide crossing, which over a wavelength range from 1520 nm to 1620 nm exhibits an insertion loss <0.08 dB (minimum 0.023 dB) and a crosstalk < -32.1 dB (minimum of < -35 dB). In addition, we also show a first test of the chemical-mechanical polishing (CMP) that we recently included in our process portfolio to enable additional photonic integration prospects, e.g., 2D material integration. In doing so, we showcase the platform's potential for a wide range of photonic applications in a wide range of wavelengths, paving the way for future, highly integrated innovations in integrated optics.

1.2 Technology

The dielectric photonic platform presented in this work is realized on 200 mm silicon wafers, featuring a 2.75 μm thick thermally grown SiO2 layer. On this base, we are able to deposit different dielectric materials, depending on the application requirements. For UV applications, we can deposit aluminum nitride (AlN) using a sputtering process that employs an aluminum target and runs at a temperature of 300 °C—well within the thermal budget for post-CMOS integration. At present, we deposited AlN layers with a thickness of 350 nm and different thicknesses can be achieved by adapting the sputtering process. For applications in the visible (VIS) and in the infra-red (IR) spectrum, we deposit different types of silicon nitride (SiN). We utilize various deposition techniques, including low pressure chemical vapor deposition (LPCVD) for a stoichiometric Si3N4, plasma enhanced PECVD Si_XN_Y and an optimized SiN for applications at visible wavelengths, PE-VIS Si_XN_Y. Up to date, we successfully deposited and characterized SiN layer thicknesses from 100 nm to 800 nm.

After depositing the dielectric material, the waveguides are defined via optical lithography, adhering to a minimum feature size of 250 nm, and subsequently etched by reactive ion etching. After etching, the photonic structures are clad by a 1 μm thick layer of undoped silicate glass (USG). Scanning electron microscopy (SEM) images of exemplary waveguide cross sections are shown in Fig. 1.1. Figure 1.1a

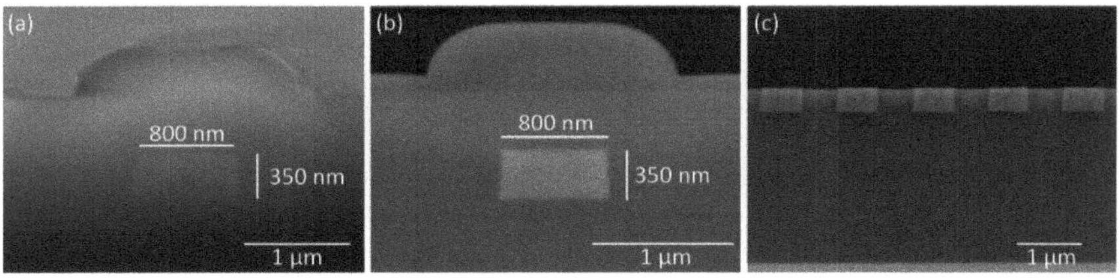

Fig. 1.1 SEM cross sections of: **a** a waveguide of sputtered AlN with a USG cladding and a sidewall angle of 75°, **b** a PECVD Si_XN_Y waveguide clad with USG with a sidewall angle of 85°, **c** an array of waveguides, where the cladding was planarized for further integration

shows an AlN waveguide, and Fig. 1.1b shows a PECVD Si_XN_Y waveguide. As a result of depositing the USG on photonic structures, the surface of the wafer is uneven, featuring bumps and valleys, see Figs. 1.1a and b. This leads to difficulties in integrating further layers, e.g., 2D materials [5]. Furthermore, an uneven surface of waveguide cores can lead to elevated scattering losses [6, 7]. To overcome these challenges, we employ chemical-mechanical polishing (CMP), which is a process that combines chemical and mechanical means to flatten the wafer surface. Figure 1.1c shows an array of PECVD Si_XN_Y waveguides which have undergone CMP.

1.3 Devices

To build upon the previously reported on feasibility of our platform, demonstrated by measuring the waveguide losses at wavelengths ranging from the blue to the infrared [4], we designed, fabricated and characterized a number of photonic devices. Here, we present a selection of these devices: a waveguide and ring resonator in AlN for the C-band, a multi-mode interferometer (MMI) for a visible wavelength, and a waveguide crossing for the C-band.

Characterization of the devices under test (DUT) in the C-band was done by sending continuous wave (CW) light from a tunable laser source via a single-mode fiber (SMF) through a grating coupler to the DUT. At the DUT output, the light was collected and sent through another grating coupler to an optical spectrum analyzer via a multi-mode fiber (MMF). At the VIS wavelengths, we used a CW laser diode emitting at 633 nm coupled to a SMF, which was used to send the light to the DUT through an edge coupler. At the DUT output, the light was collected and sent through another edge coupler to an optical power meter via another SMF.

AlN Waveguides and Rings. AlN structures were realized in a $h = 350$ nm thick photonic layer, and the designed center width of the waveguides was $w = 800$ nm, see Fig. 1.1a. We measure the attenuation by employing the cutback method. To this end waveguides of lengths ranging from 2.58 cm to 12.78 cm were patterned. In the wavelength range from 1520 nm to 1620 nm the attenuation varied between 1.93 dB/cm and 2.73 dB/cm, with the minimum attenuation near 1620 nm, see. Figure 1.2a.

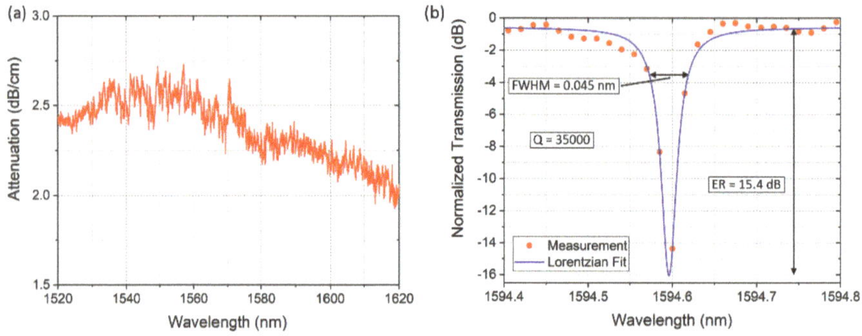

Fig. 1.2 AlN platform characterization: **a** Attenuation of 350 nm thick and 800 nm wide AlN waveguides. **b** Normalized transmission of an AlN ring resonator with a 200 μm diameter

These measured values are larger than the previously reported value of 0.16 dB/cm [8], however, our results refer to low-temperature sputtered, post-CMOS compatible AlN and are, to the best of our knowledge, the first such results.

In addition, we tested a ring resonator in this platform—the ring waveguide lateral dimensions were identical to the ones of the straight waveguide ($w \times h = 800$ nm $\times$ 350 nm), with a ring diameter of 200 μm. An excerpt of the measured spectrum is plotted in Fig. 1.2b. The relatively low Q factor is attributed to the attenuation of the waveguides.

VIS 1 x 2 Multi-mode Interferometer (MMI). The **1 x 2** MMIs that we present here were designed for use in the visible spectrum, specifically for 50/50 splitting at $\lambda = 633$ nm. They were realized in a 100 nm thick layer of our PE-VIS Si_XN_Y where each has a footprint of 142 μm^2. A microscope image of a DUT consisting of several **1 x 2** MMIs in a tree configuration is shown below, see Fig. 1.3. The measured loss per device amounted to 3.66 ± 0.04 dB. This includes losses from two 90° bends per instance. No experimental data is available for their impact, but simulations suggest a loss per 90° bend of 0.27 dB. Thus, each output branch of the MMI receives an estimated 3.12 ± 0.04 dB of the optical power entering the MMI. This shows comparable behavior to that previously shown in other work [9, 10].

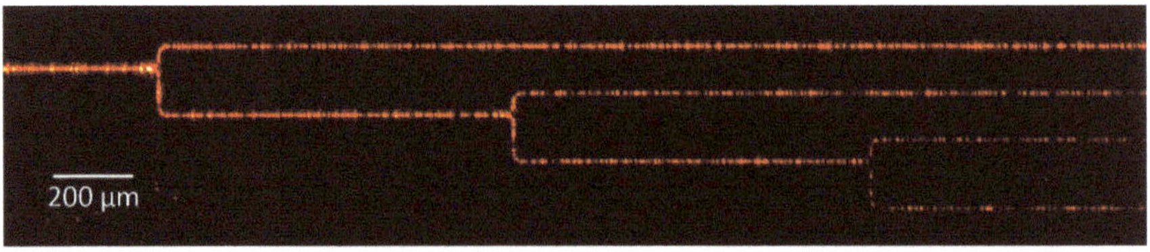

Fig. 1.3 Microscope image of a tree of 1 × 2 MMIs fabricated in the PE-VIS silicon nitride platform for a 50/50 splitting ratio. The structures were fabricated for and investigated at a wavelength of λ = 633 nm, which is traveling from left to right

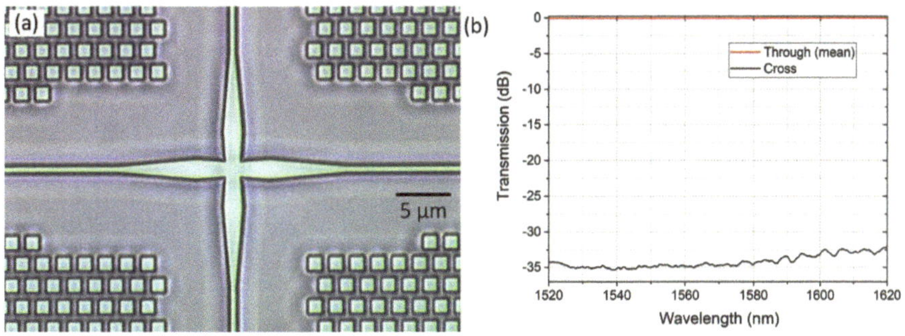

Fig. 1.4 SiN waveguide crossing for the C-band: **a** Microscope image of a waveguide crossing. **b** The transmission spectrum of the respective through and cross ports in the C-band of the same device. Through transmission was found by averaging over a large number of crossings, cross transmission is found from a single crossing

Si_XN_Y Waveguide Crossing. The waveguide crossings depicted in Fig. 1.4a were realized in our 350 nm thick PECVD Si_XN_Y platform. They consist of a taper, followed by an inverse taper towards the crossing area. This inverse taper, initially shown in [11] focusses the light within the device, minimizing crosstalk. Figure 1.4b shows the insertion loss and the crosstalk. Insertion loss was evaluated by comparing transmission from structures with 0, 20, 40, 60, 80, 100 and 130 added crossings. Crosstalk was measured from a single crossing. The minimal insertion loss was found at 1612 nm with a loss of 0.023 dB and over the whole measurement spectrum from 1520 nm to 1620 nm it never exceeded 0.08 dB. This aligns well with the corresponding simulation results, as the same boundary values for insertion loss were found, while a shift in peak performance was detected, as the lowest insertion loss was expected at 1550 nm. Crosstalk is maximal at 1620 nm with −32.1 dB and drops below −35 dB around 1540 nm, which is ~15 dB higher than what simulations suggest. Insertion loss and crosstalk figures are comparable to literature values [11].

1.4 Summary and Outlook

We presented the progress of the ongoing development of our photonic platform. We demonstrated the viability of post-CMOS compatible AlN photonic devices by low temperature AlN sputtering and subsequent structuring of photonic waveguides and ring resonators. This demonstration paves the way towards post CMOS applications in the UV. In addition, we further expand our portfolio of SiN devices. In our SiN platform optimized for the visible spectrum (PE-VIS Si_XN_Y), we fabricated and experimentally tested 1×2 multi-mode interferometers (MMIs) at a wavelength of $\lambda = 633$ nm, with excess losses comparable to the state of the art. Lastly, in the

PECVD Si_XN_Y platform, which targets applications in the C-band, we introduced waveguide crossings—experimentally measured insertion loss and crosstalk figures are on par with values found in the literature.

In the future, we are aiming to expand our photonics portfolio with 2D materials and other active materials, e.g., via atomic layer deposition or elsewise. Planarized surfaces may also act as foundation for deposition of additional photonic layers. To this end, first steps have been taken by testing the chemical-mechanical polishing (CMP) process on our SiN platform.

Acknowledgements The work presented was carried out in the framework of the APECS Pilot Line of the Chips Joint Undertaking, funded by Horizon Europe (ID: 101183307) and Digital Europe (ID: 101182906) Programms and national funding authorities of Austria, Belgium, Finland, France, Germany, Greece, Portugal, Spain. This work was supported by "terahertz.NRW". The project "terahertz.NRW" is receiving funding from the programme "Netzwerke 2021", an initiative of the Ministry of Culture and Science of the State of Northrhine Westphalia. The sole responsibility for the content of this publication lies with the authors.

This work was supported by "NANO-INC". The project "Nanolaminate für innovative nichtlineare nanophotonische Schaltkreise (NANO-INC, FKZ 13 N17104) "is funded by the Federal Ministry of Education and Research of Germany. The sole responsibility for the content of this publication lies with the authors.

This work was supported by the project "FREDERIC". The project "FouRiEr Domain ElectRIcal-to-optical Converter" is funded by the German Research Foundation (DFG)—Project number 471599135.

References

1. Shekhar, S., et al.: Roadmapping the next generation of silicon photonics. Nat. Commun. **15**, 751 (2024). https://doi.org/10.1038/s41467-024-44750-0
2. Siew, S.Y., et al.: Review of silicon photonics technology and platform development. J. Lightwave Technol. **39**, 4374–4389 (2021). https://doi.org/10.1109/JLT.2021.3066203
3. Smit, M., et al.: An introduction to InP-based generic integration technology. Semicond. Sci. Technol. **29**(8), 083001 (2014). https://doi.org/10.1088/0268-1242/29/8/083001
4. Westhues, M., et al.: Highly flexible dielectric platform for post-CMOS photonics. In: Witzens, J., Poon, J., Zimmermann, L., Freude, W. (eds.) The 25th European Conference on Integrated Optics. ECIO 2024, vol. 402. Springer Proceedings in Physics, Springer, Cham. https://doi.org/10.1007/978-3-031-63378-2_96
5. Wang, Z., et al.: Graphene-based silicon photonic devices for optical interconnects. Adv. Funct. Mater. **34**(9), 2307461 (2024). https://doi.org/10.1002/adfm.202307461
6. Payne, F.P., Lacey, J.P.R.: A theoretical analysis of scattering loss from planar optical waveguides. Opt. Quant. Electron. **26**, 977–986 (1994). https://doi.org/10.1007/BF00708339
7. Cano Castro, S., et al.: Photonic platform for industrially microfabricated ion traps. In: Witzens, J., Poon, J., Zimmermann, L., Freude, W. (eds.) The 25th European Conference on Integrated Optics. ECIO 2024, vol. 402. Springer Proceedings in Physics, Springer, Cham. https://doi.org/10.1007/978-3-031-63378-2_38
8. Singh, R., et al.: Sputtered Aluminum nitride waveguides for the telecommunication Spectrum with less than 0.16 dB/cm loss. Opt. Express. **32**, 46522–46528 (2024). https://doi.org/10.1364/OE.538852

9. Blasco-Solvas, M., et al.: Silicon nitride building blocks in the visible range of the Spectrum. J. Lightwave Technol. **42**, 6019–6027 (2024). https://doi.org/10.1109/JLT.2024.3404639
10. Sacher, W.D., et al.: Visible-light silicon nitride waveguide devices and implantable neurophotonic probes on thinned 200 mm silicon wafers. Opt. Express. **27**, 37400–37418 (2019). https://doi.org/10.1364/OE.27.037400
11. Han, L., et al.: Ultralow-loss waveguide crossing for photonic integrated circuits by using inverted tapers. Opt. Express. **30**(5), 6738–6745 (2022). https://doi.org/10.1364/OE.451104

Chapter 2
Self-configuring MZI-Based Photonic Core for Matrix-Vector Multiplication and Optical Computing

Alessandro di Tria, Pietro Giannoccaro, Marcelo Carnica, Gabriele Cavicchioli, Francesco Morichetti, Andrea Melloni, Giorgio Ferrari, Marco Sampietro, and Francesco Zanetto

Abstract We present a novel technique to control the operation of Mach-Zehnder interferometers integrated inside photonic processors. By leveraging transparent photodetectors and local feedback loops, it is possible to control the internal and output power of the device with precision as high as 11 bits. The concurrency of two local feedbacks per interferometer allows to precisely and automatically set any arbitrary 2×2 transfer matrix, enabling scalable optical computing.

2.1 Introduction

The exponential growth of high-performance computing applications such as neural networks and Artificial Intelligence (AI) is facing the limitations of traditional electronic hardware in terms of bandwidth and energy consumption. Graphics Processing Units (GPUs) and Application-Specific Integrated Circuits (ASICs) are suitable to perform highly-parallelized operations such as matrix-vector and matrix-matrix multiplications that are crucial in deep learning algorithms; however, they are intrinsically limited in latency and energy consumption per operation, mainly due to their metal interconnects [1].

A promising alternative is offered by integrated photonic platforms. Some photonic architectures can perform matrix operations entirely in the optical domain, reaching a computational speed of few ps and an energy consumption in the order of

A. di Tria (✉) · P. Giannoccaro · M. Carnica · G. Cavicchioli · F. Morichetti · A. Melloni · M. Sampietro · F. Zanetto
Department of Electronics, Information and Bioengineering, Politecnico di Milano, Milan, Italy
e-mail: alessandro.ditria@polimi.it

G. Ferrari
Department of Physics, Politecnico di Milano, Milan, Italy

P. Smowton and T. Krauss (eds.), *Proceedings of the 26th European Conference on Integrated Optics*, Springer Proceedings in Physics 440,
https://doi.org/10.1007/978-3-032-29673-3_2

fJ/operation [2]. Among them, photonic processors based on Mach-Zehnder interferometers (MZIs) are gaining popularity for their design and fabrication simplicity and for their ability to perform on-chip operations with complex numbers. However, the complexity of setting the working point of each MZI for the desired operation is hindering the development of reliable large-scale photonic processors. Indeed, the configuration of each interferometer typically requires the construction of a look-up table to link the state of each interferometer to the output of the entire processor, which can be measured by means of photodetectors [3]. This approach can be time-consuming and imprecise, since the circuit is sensible to environmental changes and requires periodic re-calibrations.

This work proposes a control technique to automatically set a unique working point for a MZI to implement a unitary 2×2 transfer matrix with high precision. The technique is calibration-free and can be scaled to high-order matrices implemented in a photonic processor with many interferometers.

2.2 Photonic Core

A balanced Mach-Zehnder interferometer with two phase shifters φ and ϑ integrated at the input and internal arms respectively (Fig. 2.1a) implements a programmable 2×2 unitary matrix, whose complex-valued coefficients depend on the phase shift values:

$$T_{MZI} = -je^{-j\vartheta/2}\begin{bmatrix} \sin(\vartheta/2) & \cos(\vartheta/2)e^{-j\varphi} \\ \cos(\vartheta/2) & -\sin(\vartheta/2)e^{-j\varphi} \end{bmatrix} \tag{2.1}$$

In order to set a specific matrix, these values need to be programmed accurately. This task is difficult to perform by only observing the device outputs, since the same output power level can be achieved with different phase shifters commands.

We thus propose a control scheme with two concurrent local feedback loops (Fig. 2.1b, c): by monitoring the optical power after each coupler of the MZI (i.e. at the output and in the internal branch of the interferometer) with on-chip sensors, it is possible to actively drive the corresponding phase shifter without ambiguity. Indeed, by naming the two input powers A_1 and A_2 and their relative phase difference χ_A, the percentage power $B_{2\%}$ at the output of the first coupler is given by:

$$B_{2\%} = \frac{1}{2} + \frac{\sqrt{A_1 A_2}}{A_1 + A_2}\sin(\chi_A + \phi) \tag{2.2}$$

where ϕ is the programmable phase shift. The same formula is valid for the output power percentage $C_{1\%}$, which depends on the internal power and phase difference and can be tuned with phase shifter ϑ. Using a well-known input and assuming negligible insertion losses, it is sufficient to monitor the optical power in positions B_2, C_1 and C_2 to implement two local feedback loops: one linking the internal power $B_{2\%}$ to the

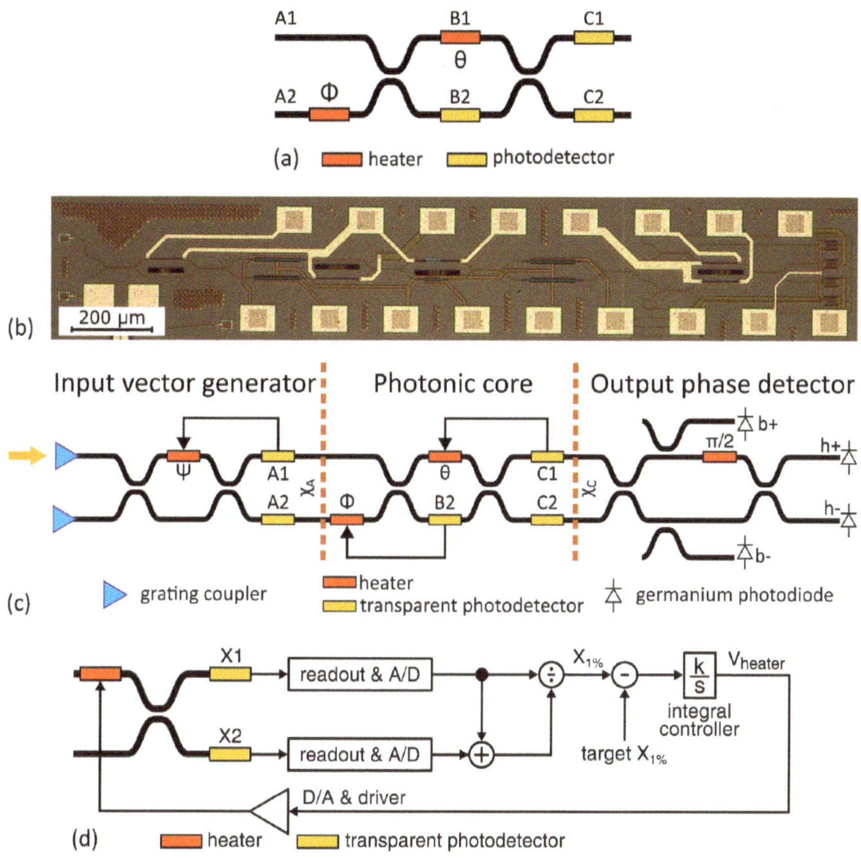

Fig. 2.1 **a** Schematic of the photonic core. The sections of the circuit have been named *A* for the input, *B* for the internal MZI arm and *C* for the output. **b** Microscope photograph and **c** schematic of the realized PIC. The arrows highlight the local feedback loops. The phase difference at the output of the core is computed as $\chi_C = \tan^{-1}\{(b_+ - b_-)/(h_+ - h_-)\}$ [4]. **d** Schematic of the electronic control loop for a generic section *X* of the MZI

actuator ϕ and one linking the output $C_{1\%}$ to the actuator ϑ. This operation uniquely sets the complex transfer matrix implemented by the interferometer, without any calibration.

2.2.1 *Practical Implementation*

We designed a Photonic Integrated Circuit (PIC) to test the control strategy. The PIC, whose photograph is shown in Fig. 2.1b, is realized in silicon photonics technology, and features thermal actuators (heaters) for the phase tuning. In order to compute the power percentages, we need to integrate sensors with a very linear response and.

which introduce minimal insertion losses, since they are placed along the optical path. We chose transversal transparent photodetectors, which satisfy both requirements [5]. These sensors rely on the propagation losses in the silicon waveguide to measure the amount of light, thus introducing no additional insertion losses.

The PIC is conceptually divided into three sections (Fig. 2.1c):

- an input vector generator, where light is injected through a grating coupler into a balanced MZI, which is controlled to generate the input vector for the successive stage;
- a photonic core, which is the MZI performing matrix-vector multiplications;
- an output phase detector [4], used to measure the phase difference at the output of the photonic core and validate the correct result of the operation.

A custom electronic system (Fig. 2.1d) has been developed to control the PIC. The system acquires signals from the on-chip photodetectors, digitizes them and implements the feedback law in the digital domain. In particular, for each section X of the interferometer it computes the power percentage $X_{1\%}$, subtracts the user-defined target value and feeds the result to an integral controller, which updates its output until the measured $X_{1\%}$ matches the target. The output of the controller is re-converted in the analog domain and used to drive the corresponding heater, effectively closing the loop in real time.

2.3 Experimental Results

2.3.1 Input Generation

The strategy has been first tested to generate the input vector using the first stage of the PIC. Light at 1550 nm is coupled through the top grating into the first MZI, where a local feedback acts on heater ψ to control the power percentage $A_{1\%}$. In order to evaluate the accuracy of the control strategy on the entire range, the target power $A_{1\%}$ has been swept from 0 to 100% with incremental steps of 0.1% every ~500 ms. The measurement is shown in Fig. 2.2a, which confirms the ability of the control to set a target optical power with high linearity on the entire range.

The resulting accuracy is also excellent, as highlighted by the histogram in Fig. 2.2b. The error distribution is computed as $A_{1\,\%\,meas} - A_{1\,\%\,target}$ and has a quasi-gaussian shape, with a zero mean and a standard deviation of only 0.046%. Using the formula $res_{acc} = \log_2(1/\sigma)$, this amounts to an accuracy resolution of 11.08 bits.

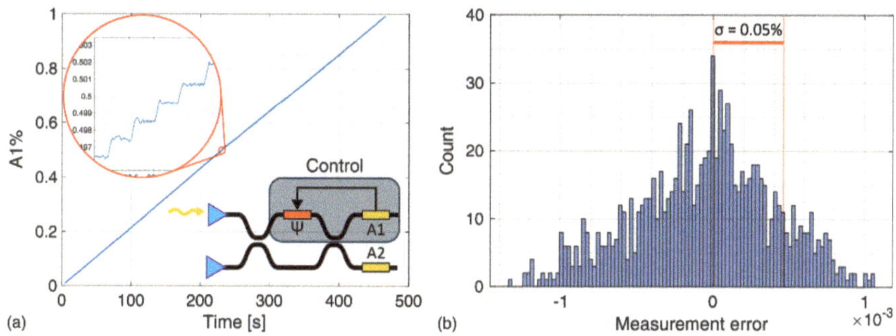

Fig. 2.2 **a** Measurement of the power percentage $A_{1\%}$ set by the local feedback, sweeping from 0 to 100% with incremental steps of 0.1% every ~500 ms. **b** Histogram of the measurement error, computed as $A_{1\ \%\ meas} - A_{1\ \%\ target}$

2.3.2 *Matrix-Vector Multiplication*

In order to test the photonic core, an input vector and a set of phase shifts φ and ϑ need to be selected. As an input, a vector of in-phase optical beams of equal power has been chosen by setting $A_{1\%} = 50\%$ with the input generator. Then, an arbitrary matrix has been selected by choosing $\varphi = 1.8\pi$ and $\vartheta = 0.15\pi$, which according to Eq. (2.2) correspond to $B_{2\%} = 20.61\%$ and $C_{1\%} = 68.36\%$, respectively.

The two local feedback loops of the photonic core have been activated to set these target phase shifts, with a response time of few tens of ms. Then, the control on the photonic core has been paused to keep the matrix fixed and the input has been changed using the input generator, in order to perform matrix-vector multiplications. A total of 9 inputs with different amplitudes and phases have been selected.

Figure 2.3 shows the output of the operations in a polar plot, where the radius is the power percentage $C_{1\%}$ measured by the output photodetectors and the angle is the phase difference χ_C measured by the output phase detector. All results are very accurate, showing a root mean square error in the measured $C_{1\%}$ power below 1% and in the measured χ_C phase of only 29 mrad, corresponding to an equivalent resolution of 8 bits. These figures of merit are better than other approaches operated without feedback [6], proving that our control strategy is effective in setting a precise target matrix without the need for pre-calibration.

2.4 Conclusions

We demonstrated a novel control technique to automatically configure a Mach-Zehnder interferometer in any desired working point. The implementation of local feedback loops with transparent photodetectors allows for precise control of the optical power in any point of the device, reaching a resolution of 11.08 bits. With

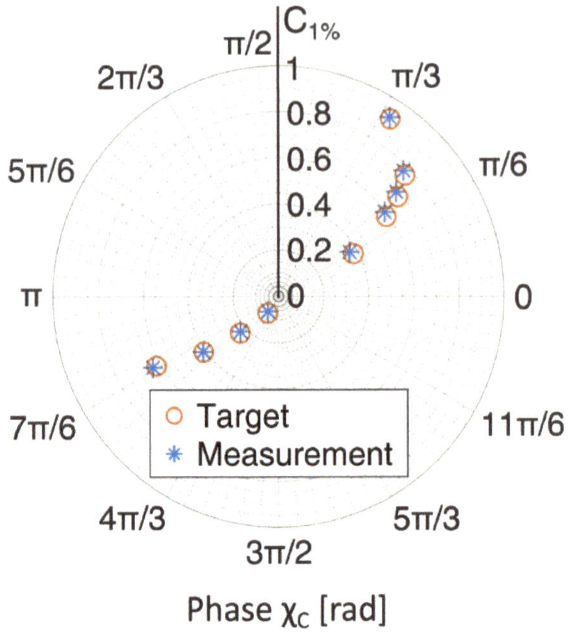

Fig. 2.3 Polar plot of the output of the photonic core in power ($C_{1\%}$) and phase difference (χ_C), for a total of 9 different inputs. The core has been configured with $\varphi = 1.8\pi$ and $\vartheta = 0.15\pi$. The measurements are indicated with blue stars, while the exact results are red circles

two local feedbacks controlling the internal and output power of the interferometer, the control is able to configure a unitary 2 × 2 transfer matrix, which can be used to perform complex-valued matrix-vector multiplications in the optical domain with 8-bit precision.

The use of transparent photodetectors and the independent configuration of the interferometer allow seamless scaling to photonic processors of any size, where many MZIs arranged in different architectures [7] can perform high-order matrix-vector multiplication.

References

1. McMahon, P.L.: The physics of optical computing. Nature Reviews Physics. **5**(12), 717–734 (2023)
2. Zhou, H., et al.: Photonic matrix multiplication lights up photonic accelerator and beyond. Light: Science and Applications. **11**(1), 30 (2022)
3. Bandyopadhyay, S., et al.: Hardware error correction for programmable photonics. Optica. **8**, 1247–1255 (2021)
4. Sun, Z., et al.: Scalable low-latency optical phase sensor array. Optica. **10**, 1165–1172 (2023)
5. Perino, A., et al.: High-sensitivity transparent photoconductors in voltage-controlled silicon waveguides. Opt. Lett. **47**, 1327–1330 (2022)
6. Rahimi, K.S., et al.: Realization of an integrated coherent photonic platform for scalable matrix operations. Optica. **11**(4), 542–551 (2024)
7. Mojaver, K.H.R., et al.: Addressing the programming challenges of practical interferometric mesh based optical processors. Opt. Express. **31**(15), 23851–23866 (2023)

Chapter 3
Towards Highly-Confined > 1 μm Thick Silicon Nitride Waveguides for Photonic Integrated Circuits

Yasser Pordeli, Francisus H. B. Somhorst, Jakub Sadilek, Kees van der Zouw, Melissa Goodwin, Erwin J. W. Berenschot, Roald M. Tiggelaar, and Pepijn W. H. Pinkse

Abstract Stoichiometric Silicon nitride (S_3N_4) is an important material for photonic integrated circuits (PICs) due to its wide transparency range, low propagation loss, and CMOS compatibility. However, the fabrication of highly-confined waveguides with thicknesses exceeding 1 μm remains challenging due to stress-induced cracking in conventional silicon-based substrates. To address this, we demonstrate a novel approach utilizing single-step low-pressure chemical vapor deposition (LPCVD) method to fabricate ~1.3 μm thick S_3N_4 waveguides with symmetric silicon dioxide (SiO_2) cladding on sapphire substrates. Our approach leverages sapphire's mechanical properties to mitigate residual stress, allowing for the deposition of thick Si_3N_4 layers without cracking. The fabrication process includes LPCVD of ~1 μm SiO_2 cladding on sapphire substrate, single-step LPCVD of ~1.3 μm Si_3N_4, subsequent etching and lithography steps to define the waveguide structures, and the second SiO_2 cladding deposition. Characterization techniques such as ellipsometry and focused ion beam scanning electron microscopy (FIB-SEM) confirm the successful deposition and structuring of the waveguides. Despite minor tapering in the waveguide profiles, our results represent a significant advancement toward scalable high-confinement Si_3N_4 waveguides. Future work will focus on optimizing plasma etching to achieve vertical sidewalls, reducing surface roughness, and evaluating optical performance. This study lays the foundation for next-generation nonlinear optics, mid-infrared applications, and high-performance photonic circuits.

Y. Pordeli (✉) · F. H. B. Somhorst · J. Sadilek · K. van der Zouw · M. Goodwin · E. J. W. Berenschot · R. M. Tiggelaar · P. W. H. Pinkse
MESA+ Institute, University of Twente, Enschede, Netherlands
e-mail: y.pordeli@utwente.nl

P. W. H. Pinkse
e-mail: p.w.h.pinkse@utwente.nl

P. Smowton and T. Krauss (eds.), *Proceedings of the 26th European Conference on Integrated Optics*, Springer Proceedings in Physics 440,
https://doi.org/10.1007/978-3-032-29673-3_3

3.1 Introduction

Silicon (Si) photonics has garnered significant interest over the last decade due to its applications in data communications and telecommunications. Among the materials for integrated optics, stoichiometric silicon nitride (Si_3N_4) stands out for its exceptional properties. It features a broad transparency range, low propagation loss, high refractive index, and is CMOS compatible [1–4]. These characteristics make Si_3N_4 an excellent candidate for non-linear optical applications. However, achieving anomalous dispersion requires highly confined waveguides with thicknesses exceeding 700 nm. Similarly, mid-infrared (mid-IR) applications demand layers thicker than 300 nm [5, 6].

Conventional fabrication of low-loss Si_3N_4 waveguides involves low-pressure chemical vapor deposition (LPCVD) on oxidized Si substrates. However, the process is constrained by the high residual tensile stress (typically >1 GPa), which limits the deposited layer thickness to 250–400 nm, as thicker layers are prone to cracking [7, 8]. These challenges hinder the performance and scalability of Si_3N_4-based photonic devices, especially for highly confined waveguides and anomalous dispersion engineering.

To overcome the thickness limitations caused by stress-induced cracking in Si_3N_4 layers on silicon substrates, several alternative fabrication approaches have been explored to increase the deposited thickness up to ~900 nm. These include, among others, sequential LPCVD depositions with intermediate cooling steps to reduce stress accumulation [9]. Another method involves pre-etching trenches into the Si substrate to act as crack terminators during deposition [10]. To simplify fabrication and improve scalability, a more recent approach demonstrated direct deposition of ~1.35 μm thick Si_3N_4 on sapphire [11]. The residual stress in the Si_3N_4 layer on sapphire was reported to be significantly lower at ~120 MPa [11]. This highlights how sapphire's mechanical properties effectively suppress stress-induced cracking, enabling the deposition of thick Si_3N_4 layers in a single LPCVD step.

Here, we demonstrate the successful fabrication of single-step LPCVD of ~1.3 μm thick Si_3N_4 with symmetric silicon dioxide (SiO_2) as cladding layers, as illustrated in Fig. 3.1. To our best knowledge, this work is the first to implement symmetric SiO_2 cladding in thick Si_3N_4 waveguides on sapphire, representing a significant advancement in integrated photonics. These findings pave the way for improved device performance in non-linear, mid-infrared, and visible light photonics integrated circuits (PICs).

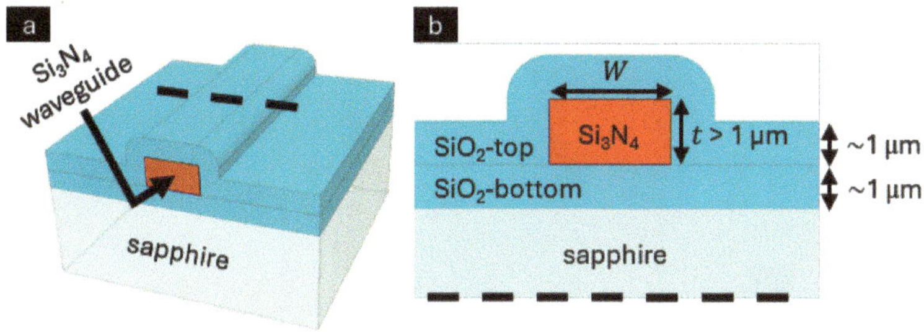

Fig. 3.1 **a** Envisioned thick stoichiometric silicon nitride (Si_3N_4) waveguide with symmetric silicon dioxide (SiO_2) cladding on a sapphire substrate. **b** Cross-sectional view indicated by the dashed black line

3.2 Experiment and Characterization

A schematic overview of the fabrication steps of thick (>1 μm) stoichiometric silicon nitride (Si_3N_4) on sapphire with symmetric silicon dioxide (SiO_2) cladding is illustrated in Fig. 3.2. For our experiments sapphire substrates (Fig. 3.2a) were selected (100 ± 0.25 mm diameter, 500 ± 25 μm thick, Kyocera Fineceramics Europe GmbH).

First, the substrates were standard pre-furnace cleaned for 40 min using ozone steam cleaning (Semozone 090.2 LP Ozone Generator, home-built at MESA+). Low-pressure chemical vapor deposition (LPCVD) was performed (Fig. 3.2b) at 725 °C to deposit a ~ 1 μm tetraethyl orthosilicate (TEOS) layer (40 sccm TEOS, 30 sccm N_2, 400 mTorr, 1 h 57 min). Next, the substrates were post-annealed at 1150 °C for 3 h to convert TEOS to silicon dioxide (SiO_2), resulting in densification of the layer and elimination of hydrogen-related bonds. Next, a single step LPCVD (22 sccm SiH_2Cl_2, 66 sccm NH_3, TS6604, 200 mTorr, Tempress) was carried out (Fig. 3.2c) at 800 °C for 5 h 10 min to deposit ~1.3 μm stoichiometric silicon nitride (Si_3N_4). Subsequently, amorphous silicon (a-Si), serving as hard mask for plasma etching, was deposited (Fig. 3.2d) by LPCVD at 550 °C to obtain a layer thickness of about 400 nm (50 sccm SiH_4, 250 mTorr, 2 h 40 min). Ellipsometry (J.A. Woollam M-2000UI) was performed to measure the thickness of the layers on sapphire substrate. Figure 3.3 shows the measured thickness at five locations on the substrate. The average and standard deviations were calculated for the 5-points, i.e. 976 ± 7 nm SiO_2, 1344 ± 20 nm Si_3N_4, and 403 ± 5 nm a-Si.

Next, positive tone photoresist (PR) of ~1.7 μm (Olin OiR 907–17. FujiFilm, Japan) was spin-coated at 4000 rpm for 30 s and pre-exposure baked at 90 °C for 90 s. Afterwards, the substrates were loaded in the maskless aligner (MLA, Heidelberg Instruments) to write straight and spiral wave guide structures (375 nm laser, 140 mJ/cm^2 dose). Then, the substrates were post-exposure baked for 60 s at 120 °C and (Fig. 3.2e) developed for 60 s in a tetramethylammonium hydroxide (TMAH) solution (OPD4252, Arch Chemicals).

The substrates are then transferred to an inductively coupled reactive ion etching (ICP-RIE) system (Oxford Plasma Pro 100 Estrelas) for anisotropic etching

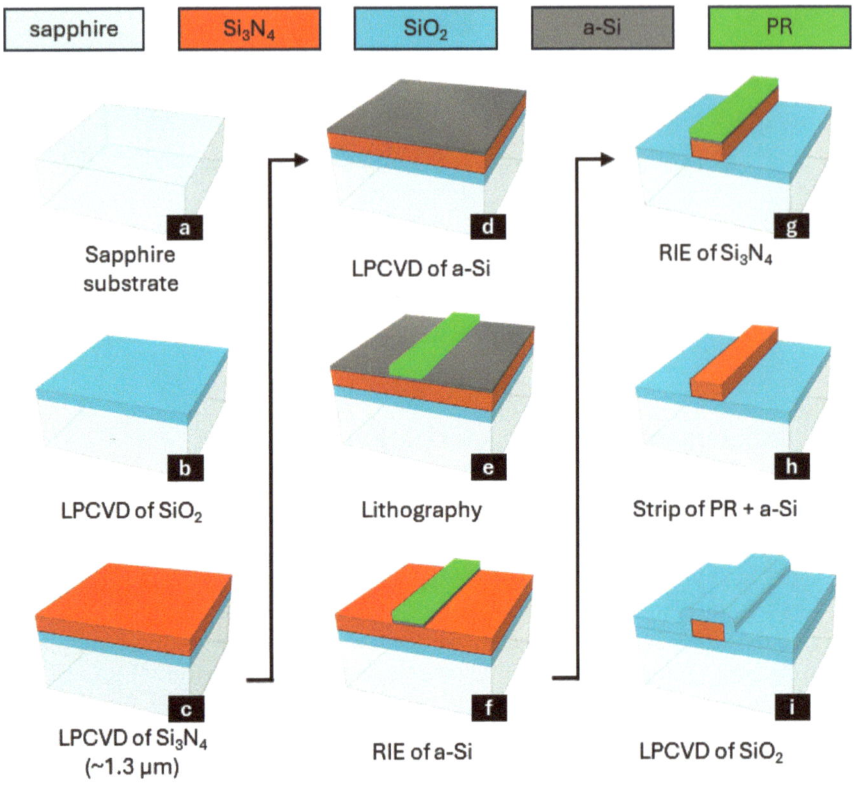

Fig. 3.2 Schematic fabrication flow of single-step low-pressure chemical vapor deposition (LPCVD) of stoichiometric silicon nitride (Si_3N_4) with symmetric silicon dioxide (SiO_2) cladding on a sapphire substrate. Amorphous silicon (a-Si) and photoresist (PR) were used as hard and soft masks, respectively. Reactive ion etching (RIE) was employed to pattern both a-Si and Si_3N_4 layers

(Fig. 3.2f) of the a-Si hard mask. A continuous mixed process using sulphur hexafluoride (SF_6) for etching and octafluorocyclobutane (C_4F_8) for sidewall passivation ensures smooth, vertical sidewalls of Si (18 mTorr, 48 sccm C_4F_8, 23 sccm SF_6, 41 W CCP, 800 W ICP, 0 °C substrate, 5 min) [12]. Next, RIE was performed (Fig. 3.2g) to etch Si_3N_4 (100 sccm CHF_3, 9 sccm O_2, 40 mTorr, 250 W, Plasmatherm PT790 parallel plate etcher), at a pre-determined etch rate of about 50 nm/min. After etching, fluorocarbon residues and PR were removed by oxygen (O_2) plasma cleaning for 5 min (100 sccm O_2, 10 mTorr, 2000 W ICP, 0 °C, Oxford Plasma Pro 100 Estrelas) followed by additional O_2 plasma (500 sccm O_2, 800 W) for 30 min (TePla 300, PVA TePla AG, Germany) and RCA-2 cleaning for 15 min (HCl:H_2O_2:H_2O at 1:1:5, 80 °C).

Afterwards, the a-Si hard mask was selectively removed (Fig. 3.2h) by first etching the native SiO_2 in hydrofluoric acid (HF) solution (1 wt.%, room temperature) for 20 s followed by etching in TMAH solution (25 wt.%, 70 °C) for 2 min 50 s. Next, the substrates were pre-furnace cleaned and second TEOS cladding layer was deposited

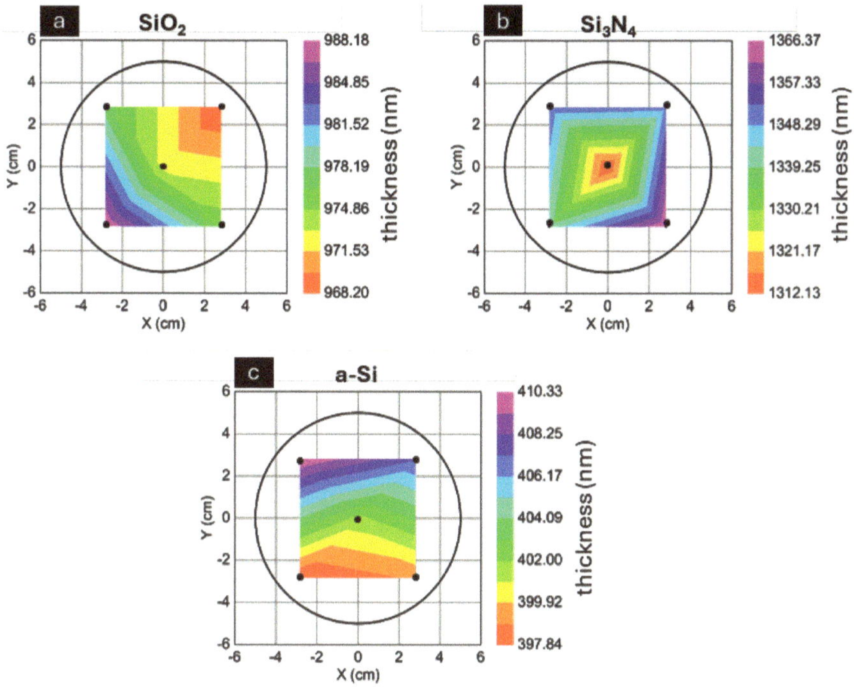

Fig. 3.3 Ellipsometry measurements of the SiO_2/Si_3N_4/a-Si stack on sapphire substrate. The circle indicates the 100-mm diameter sapphire substrate and the black dots are the measured locations

(Fig. 3.2i) by LPCVD. Subsequently, the substrates were annealed at 1150 °C for 3 h to convert the TEOS layer to SiO_2. Finally, the samples were diced in sample sizes of 10 mm by 10 mm (sapphire blade VT07 SD280 VC100 100–52 × 0.4A3 × 40–L, Disco DAD3220 dicer). The samples were manually cleaved using a diamond scriber and inspected by scanning electron microscopy (SEM), as shown in Fig. 3.4.

Although sapphire is a hard single-crystal material, it does not cleave as effectively with a diamond scriber as Si substrates do. Hence, a focused ion beam (FIB) equipped in a SEM system was used to prepare a lamella for cross-sectional inspection. Figure 3.5 shows layer stack of sapphire/$SiO_2/Si_3N_4/SiO_2$ at the final fabrication step. Despite the Si_3N_4 profile having a taper angle of approximately 81°, the fabrication was accomplished successfully. Further investigation is required to develop a recipe for the RIE of Si_3N_4 to obtain straighter profile to enhance the waveguide performance.

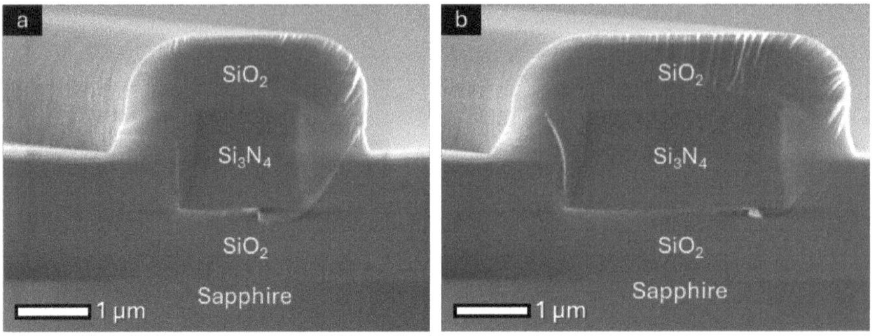

Fig. 3.4 SEM cross-sectional images **a, b** of the single-step LPCVD Si_3N_4 with symmetric SiO_2 cladding after manually cleaving with diamond scriber. Note the different widths of the fabricated Si_3N_4 waveguides in **a** and **b**

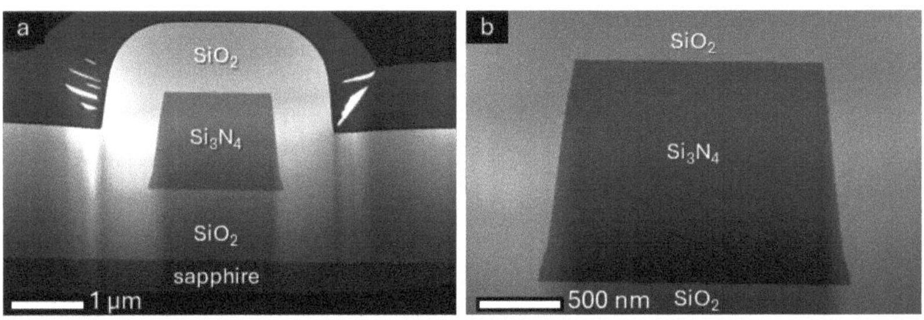

Fig. 3.5 **a** FIB-SEM cross-sectional view of the fabricated single-step LPCVD Si_3N_4 with symmetric SiO_2 cladding on sapphire substrate. **b** Close-view of the Si_3N_4 core

3.3 Summary and Outlook

In this work, we successfully demonstrated the fabrication of ~1.3 µm thick stoichiometric silicon nitride (Si_3N_4) by a single-step LPCVD on sapphire with symmetric silicon dioxide (SiO_2) cladding layer of approximately 1 µm. This fabrication platform unlocks new possibilities for advanced photonic applications, particularly in anomalous dispersion engineering and mid-IR technologies.

Several process challenges remain to be addressed to further optimize the fabrication process. The selectivity of the amorphous silicon (a-Si) hard mask with respect to the Si_3N_4 during plasma etching requires optimization. The use of a hard mask facilitates deeper etch depths and ensures compatibility with various lithography techniques that require different resists for patterning. Furthermore, the fabricated Si_3N_4 waveguide profile exhibits tapering (~ 81°), indicating that the plasma etching process requires optimization to achieve vertical sidewalls. Additionally, post-RIE fluorocarbon residue removal and sidewall roughness reduction need to be investigated to understand the effects on the performance of the waveguides.

Future work will focus on performing detailed low-loss optical measurements, determining the stresses on the layer stack, and further increasing the Si_3N_4 thickness to determine the upper limit before cracking occurs in the sapphire-SiO_2 stack.

Acknowledgements This work was made possible through funding provided by PhotonDelta. The authors gratefully acknowledge the MESA+ NanoLab staff for their continuous efforts in maintaining high standards in the facility, with special recognition to Christiaan Bruinink for his valuable support in developing the model for the ellipsometry measurements.

References

1. Roeloffzen, C.G.H., et al.: Low-loss Si_3N_4 TriPleX optical waveguides: technology and applications overview. IEEE J. Sel. Top. Quantum Electron. **24**(4), 1–21 (2018)
2. Li, X., Barbosa, F.A.S., Roberts, S.P., et al.: Ultra-low-loss on-chip resonators with sub-milliwatt parametric oscillation threshold. Optica. **4**(6), 619–624 (2017)
3. Ji, X., Jang, J.K., Dave, U.D., et al.: Exploiting ultralow loss multimode waveguides for broadband frequency combs. Laser Photonics Rev. **15**(1), 2000353 (2021)
4. Liu, K., Jin, N., Cheng, H., et al.: Ultralow 0.034 dB/m loss wafer-scale integrated photonics realizing 720 million Q and 380 μW threshold Brillouin lasing. Opt. Lett. **47**(7), 1855–1858 (2022)
5. Krückel, C.J., Fülöp, A., Ye, Z., Andrekson, P.A., Torres-Company, V.: Optical bandgap engineering in nonlinear silicon nitride waveguides. Opt. Express. **25**(13), 15370–15380 (2017)
6. Cui, S., et al.: Compact microring resonator based on ultralow-loss multimode silicon nitride waveguide. Adv. Photonics Nexus. **2**(4), 046007 (2023)
7. Hong, S., Weihs, T.P., Bravman, J.C., et al.: Measuring stiffnesses and residual stresses of silicon nitride thin films. J. Electron. Mater. **19**(9), 903–909 (1990)
8. Li, Z., et al.: Process development of low-loss LPCVD silicon nitride waveguides on 8-inch wafer. Appl. Sci. **13**(6), 3660 (2023)
9. Gondarenko, A., Levy, J.S., Lipson, M.: High confinement micron-scale silicon nitride high Q ring resonator. Opt. Express. **17**(14), 11366–11370 (2009)
10. Luke, K., Dutt, A., Poitras, C.B., et al.: Overcoming Si_3N_4 film stress limitations for high quality factor ring resonators. Opt. Express. **21**(19), 22829–22833 (2013)
11. Martinussen, S., et al.: Thick waveguides of low-stress stoichiometric silicon nitride on sapphire (SiNOS). Opt. Express. **32**(21), 36835–36847 (2024)
12. Jonker, D., et al.: Wafer-scale fabrication and modification of silicon nanopillar arrays for nanoelectronics, nanofluidics, and beyond. Int. J. Nanotechnol. **17**, 583 (2020)
13. Muñoz, P., Micó, G., Bru, L.A., et al.: Silicon nitride photonic integration platforms for visible, near-infrared, and mid-infrared applications. Sensors. **17**(9), 2088 (2017)

Chapter 4
A Compact High-Q Microdisc Resonator for Widely Tunable Microwave Photonic Filters

Suvarna Parvathy, Pawan Kumar Pandit, and Bijoy Krishna Das

Abstract A compact high-Q silicon photonic micro-disc resonator design is proposed and demonstrated for energy-efficient tunable microwave photonic filters. The disc radius of 5-μm is carefully chosen such that its fundamental whispering gallery mode is phase-matched with the guided fundamental mode of its bus waveguide of width 540 nm, which was fabricated in a silicon-on-insulator (SOI) substrate of device layer thickness 220 nm. A spiral Tungsten microheater (cross-section 600 nm $\times$ 400 nm) of length 103-μm was integrated on the top cladding layer for an efficient thermo-optic tuning. The Q-value of the fabricated resonators are as high as $\sim 2 \times 10^5$ for a resonance wavelength $\lambda_r \sim 1550$ nm corresponding to the fundamental whispering gallery mode. The fabricated microdisc resonator devices were successfully used for the demonstration of a widely and continuously tunable microwave filter (tuning range of up to 50 GHz and filter bandwidth of $\sim$ 1 GHz with a thermo-optic tuning efficiency as high as $\sim$ 38 GHz/mW).

4.1 Introduction

Chip-scale silicon photonic microwave photonic filter designs have been popular in recent times because of their wide range of frequency tunability and reconfigurability as well as CMOS fabrication process compatibility. While high-Q microring resonators have been widely investigated for the demonstration of various microwave photonic filter functions [1–3], microdisc resonators have only been reported a few [4–6]. Nevertheless, a compact design of microdisc resonator (radius $\sim 5 - \mu$m) can be thermo-optically tuned efficiently over a broad range of wavelength (FSR $\sim$ 20 nm) without mode-hopping. However, the major challenge is the fabrication of a compact and high-quality microdisc resonator. In this work, we have exploited a controlled coupling between the fundamental whispering gallery mode of the disc of 5-μm radius and the guided mode of its single-mode bus waveguide to demonstrate

S. Parvathy (✉) · P. K. Pandit · B. K. Das
Centre for Programmable Photonic Integrated Circuits and Systems, Department of Electrical Engineering, IIT Madras, Chennai, India
e-mail: ee18d416@smail.iitm.ac.in

P. Smowton and T. Krauss (eds.), *Proceedings of the 26th European Conference on Integrated Optics*, Springer Proceedings in Physics 440,
https://doi.org/10.1007/978-3-032-29673-3_4

reasonably high-Q value resonances at $\lambda_r \sim 1550$ nm. The fabricated microdisc resonators have been used further to demonstrate thermo-optically tunable microwave photonic filters. The experimental results have been found to be a good match with theoretical prediction. In Sect. 4.2, we have presented the design and characterization of the proposed thermo-optically tunable micro-disc resonator. The experimental demonstration of widely tunable microwave photonic filter function has been discussed in Sect. 4.3. Finally, a brief concluding remarks have been made in Sect. 4.4.

4.2 Device Design and Characteristics

We first investigated whispering gallery modes of a silicon photonic microdisc of 5 μm radius to operate in all-pass configuration. The device layer of the silicon-on-insulator (SOI) substrate was assumed 220-nm so that it could be fabricated using a commercial silicon photonics foundry through multi-project-wafer runs (IMEC iSiPP50G). The calculated effective index of the fundamental whispering gallery mode is 2.5041 ($\lambda \sim 1550$ nm). Accordingly, the single-mode bus waveguide width was fixed to $\sim$ 540 nm, such that its guided mode (TE_0) is exactly phase-matched with that of fundamental whispering gallery mode (see Fig. 4.1a). The bus waveguide and disc gap of about 150 nm (critical dimension for the foundry process) offers an estimated Q-value of $\sim 3 \times 10^5$ at a resonance wavelength $\lambda_r \sim 1550$ nm, assuming typical propagation loss of 2 dB/cm for the fundamental whispering gallery mode (see Fig. 4.1b).

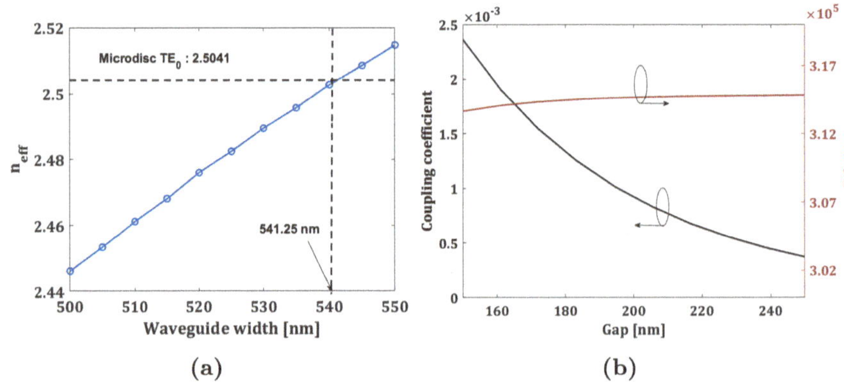

Fig. 4.1 **a** Calculated effective index of the fundamental guided mode as a function bus waveguide width (strip waveguide). Dotted line indicates the effective index of the fundamental whispering gallery mode of a 5-μm silicon photonic disc resonator; **b** Coupling coefficient and Q factor (loss = 2 dB/cm) as a function of gap between bus waveguide (width 540 nm) and disc resonator (radius 5 μm) for the fundamental whispering gallery mode (TE_0)

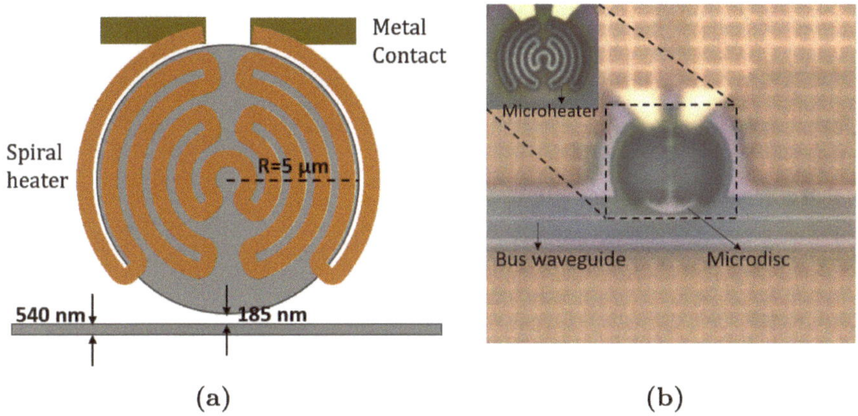

Fig. 4.2 **a** Schematic of the microdisc resonator with spiral microheater; **b** Microscopic image of microdisc resonator. Inset shows zoomed and focused image of the spiral microheater

A few different gap variations (160 nm, 185 nm and 210 nm) were considered for fabrication based on the simulation results. The device was fabricated along with a spiral Tungsten microheater (cross-section 600 nm × 400 nm) of length 103-μm integrated in the top cladding layer for an efficient thermo-optic tuning, following foundry defined fabrication process rules (see Fig. 4.2). The heater was placed with a lateral distance of around 2 μm from the coupler region to ensure that coupling coefficient is not disturbed by heating. Figure 4.2a shows the schematic of the MDR and 2b shows the microscopic image of the fabricated device, with the dimensions as specified in 2a. The inset in Fig. 4.2b shows the magnified and focused image of the spiral heater. The transmission characteristics in all pass configuration exhibits resonances corresponding to 0th order (TE_0), 1st order (TE_1) whispering gallery modes as shown in Fig. 4.3a. The free spectral range (FSR) is 20.67 nm and 21.61 nm for the fundamental and first order modes respectively. The inset shows the zoomed image of a resonance around 1550 nm corresponding to TE_0 mode. The non-Lorentzian shape of the resonance is due the presence of non-linearities. The estimated Q-value for a resonance corresponding to TE_0 mode around 1550 nm is $\sim 1.72\times10^5$. Thermo-optic tuning characteristics using the spiraled heater is also shown in Fig. 4.3b for both the modes and they are identical. The power required to tune half the FSR (P_π) is coming to be 29.37 mW, with a tuning efficiency of ~351 pm/mW.

4.3 Experimental Demonstration of Microwave Photonic Filters

A micro-disc resonator described above has been used to demonstrate tunable microwave photonic filter using the experimental setup as schematically shown in Fig. 4.4a. The optical carrier from a laser source operating at $\lambda_c = 1548.738$ nm close

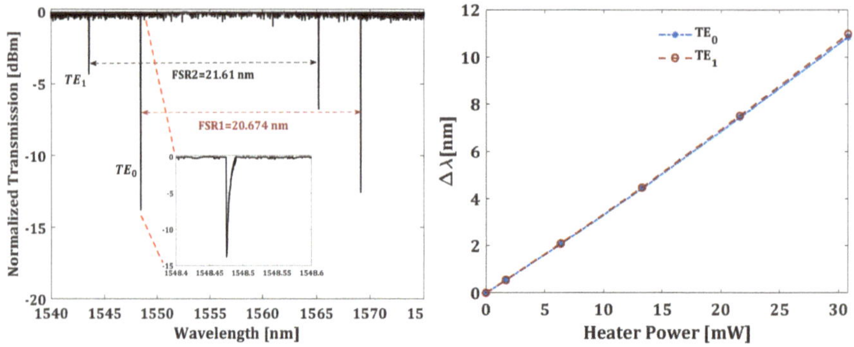

(a) (b)

Fig. 4.3 **a** Transmission characteristics of fabricated MDR (bus-to-disc gap of 185 nm) operating at room temperature. The inset shows the zoomed image of a resonance around 1550 nm; **b** Resonance wavelength detuning ($\Delta\lambda$) as a function of heater power for TE_0 and TE_1 whispering gallery modes

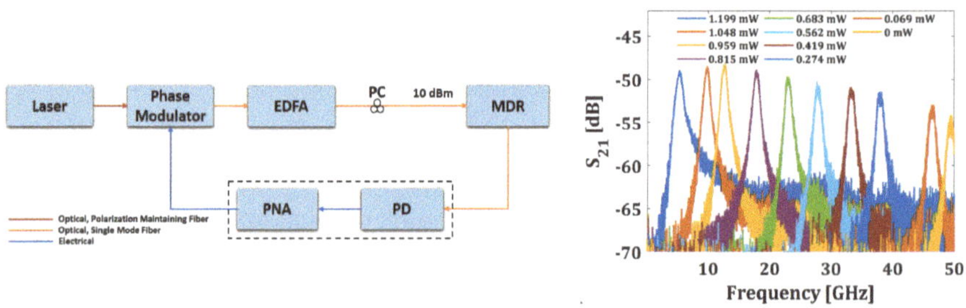

(a) (b)

Fig. 4.4 **a** Experimental setup for microwave filter demonstration. EDFA—Erbium doped fiber amplifier, PC—Polarization controller, MDR—Microdisc resonator, PD—Photodetector, PNA—Performance network analyser; **b** Microwave bandpass filter response with centre frequency tuning up to 50 GHz

to a resonance wavelength of λ_r = 1548.342 nm (fundamental whispering gallery mode) is passed through a broadband $LiNbO_3$ phase modulator (PM) and then launched into the micro-disc resonator. Finally, the filter performance is obtained at the output of broadband photodetector (PD) feeding into a performance network analyzer (PNA). The modulator is driven by sinusoidal RF frequencies sourced from the same PNA. Further details of the experimental setup may be found elsewhere [2].

Figure 4.4b shows the filter response with centre frequency tuned from 5 GHz to 50 GHz (by tuning the resonance wavelength λ_r) which is primarily limited by the measurement systems. The drop in the gain of filter as frequency increases is due to the bandwidth limitation of phase modulator (40 GHz). The heater power consumption is only 1.2 mW to tune from 5 to 50 GHz, which gives a tuning efficiency of 37.8 GHz/mW. The bandwidth of the filter is around 1–1.3 GHz. This is

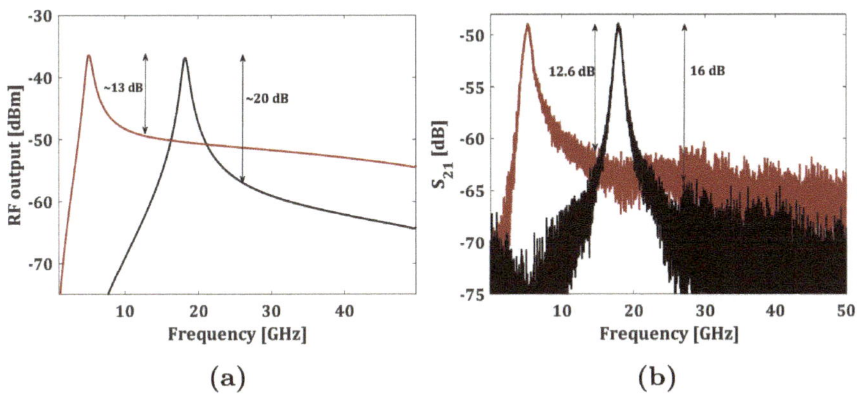

Fig. 4.5 **a** Simulated filter response around 5 GHz and 18 GHz; **b** Experimental filter response around 5 GHz and 18 GHz, respectively

the best reported energy-efficient tuning performance to the best of our knowledge. As seen from the frequency response, for lower centre frequencies f_c, the sideband rejection is asymmetric because of the residual phase response of the MDR, as discussed in [3]. However, as f_c increases, the rejection tends to be more symmetric as the influence of MDR phase fades out. The simulated results exhibiting RF filter response at 5 GHz and 18 GHz are shown in Fig. 4.5a, where the sideband rejection is highly asymmetric at 5 GHz and becomes more symmetric around 18 GHz. The asymmetricity in side-band rejection is directly related to the phase influence on optical carrier wavelength λ_c due to its proximity to the resonance wavelength λ_r. The experimental results corresponding to these frequencies are shown in Fig. 4.5b. We get a side-band rejection of about 12.6 dB and 16.0 dB, respectively for 5 GHz and 18 GHz, reaching up to 20 dB on the lower frequency range. The experimental results are closely matching with the simulation results.

4.4 Summary and Conclusion

We have demonstrated a widely tunable microwave photonic filter using a compact micro-disc resonator with Q-value as high as 1.7×10^5. The microwave band-pass filter response with a typical bandwidth of $\sim$ 1 GHz could be tuned thermo-optically upto 50 GHz with a tuning slope as high as $\sim$ 38 GHz/mW. Table 4.1 compares the performance figures-of-merit of the microwave photonic filters demonstrated using silicon photonic micro-disc resonators which have been reported till date in the published literature (to the best of our knowledge) along with results demonstrated

Table 4.1 Performance comparison of microdisc-based microwave photonic filter demonstrated in this paper with those of earlier reported works

Disc radius (μm)	Q factor	Maximum tuning range (GHz)	Filter bandwidth (GHz)	Tuning efficiency	Sideband rejection (dB)	Reference
10	8.5×10^4	10	1.93	8.56 GHz/mW	13.0	[4]
40	2.74×10^6	30	0.08	NA	27.3	[5]
4	7.8×10^4	14	3.00	NA	15.0	[6]
5	1.72×10^5	50	1.00	37.8 GHz/mW	16.0	This work

in this paper. It is evident that the proposed microring resonator design is superior in terms of thermo-optic tuning efficiency and tuning range, thus making it suitable for chip-scale microwave photonic circuit applications.

Acknowledgements The authors gratefully acknowledge MeitY, Government of India for the establishment of Silicon Photonics CoE-CPPICS at IIT Madras.

References

1. Zhang, W., Jiang, Y., Liu, Y., Wang, B.: Fully tunable microwave photonic narrow bandpass filter using an on-chip dual-drive microring resonator. Opt. Express **32**, 31043–31054 (2024)
2. Velamuri, A., Das, B.: Programmable silicon photonic RF filters with symmetric out-of-band rejection. J. Lightwave Technol. **42**, 1586–1596 (2023)
3. Li, J., Zheng, P., Hu, G., Zhang, R., Yun, B., Cui, Y.: Performance improvements of a tunable bandpass microwave photonic filter based on a notch ring resonator using phase modulation with dual optical carriers. Opt. Express **27**, 9705–9715 (2019)
4. Zhang, W., Yao, J.: On-chip silicon photonic integrated frequency-tunable bandpass microwave photonic filter. Opt. Lett. **43**, 3622–3625 (2018)
5. Cheng, Y., Hong, X., Wang, B., Zhang, W.: Widely tunable narrowband integrated microwave photonic filter based on an ultra-high-Q micro-disk resonator on silicon. In: 27th OptoElectronics and Communications Conference (OECC) and 2022 International Conference on Photonics in Switching and Computing (PSC), pp. 1–3 (2022)
6. Zhang, W., Yao, J.: Silicon-based on-chip microdisk resonators for integrated microwave photonic applications. Optical Fiber Communication Conference, pp. M2B–6 (2016)

Chapter 5
Multiphysics Simulations on the Frequency Response of Ultrasonic Detectors Based on Polymer Waveguide Microring Resonators

Rainer Hainberger, Jan-Jonas Schumacher, Paul Muellner, and Moritz Eggeling

Abstract Integrated optical waveguide microring resonators (MRRs) are promising ultrasonic detectors for photoacoustic imaging and other applications due to their compact size, high sensitivity, and potential for high frequency response and array configurations. Polymer-based MRRs are particularly advantageous as they can be fabricated on transparent substrates, enabling optical excitation of the ultrasonic shockwave through the substrate. A thin polymer cladding on the polymer waveguide MRR provides mechanical protection and can enhance the ultrasonic detection sensitivity.

This study uses a 2D FEM multiphysics simulation model implemented in COMSOL to analyze the dynamic optical response of polystyrene waveguide MRRs covered with a PDMS cladding to acoustic waves. We use the effective refractive index shift induced by the acoustic waves to assess this response, examining the influence of cladding thickness, elasticity, and elasto-optic properties as a function of the acoustic frequency. This analysis provides deeper physical insights in the mechanisms governing the optical response of polymer waveguide MRRs with cladding. The results underline that the Possion's ratio of the cladding significantly affects the frequency response, potentially limiting the ultrasonic detection bandwidth to below 10 MHz. At certain higher frequencies, standing acoustic waves can form in the cladding. Since the Possion's ratio of PDMS strongly varies with the processing conditions, precise control over fabrication is essential. In conclusion, this study highlights the crucial role of the elastic properties of the cladding on MRR ultrasonic detectors.

R. Hainberger (✉) · J.-J. Schumacher · P. Muellner · M. Eggeling
AIT Austrian Institute of Technology GmbH, Vienna, Austria
e-mail: rainer.hainberger@ait.ac.at

P. Smowton and T. Krauss (eds.), *Proceedings of the 26th European Conference on Integrated Optics*, Springer Proceedings in Physics 440,
https://doi.org/10.1007/978-3-032-29673-3_5

5.1 Introduction

Integrated optical waveguide microring resonators (MRR) have attracted strong interest as ultrasonic detectors for photoacoustic imaging and other applications because they are compact and highly sensitive [1]. Moreover, they can offer high bandwidth and the potential for array configurations. Among the different types of MRRs used as ultrasonic detectors in terms of waveguide material systems and transduction.

mechanisms, polymer-based concepts are playing a strong role. One advantage of polymer waveguides is that they can be easily fabricated on transparent substrates, which is a prerequisite for many photoacoustic imaging applications in order to allow optical excitation of the ultrasonic shockwave through the MRR substrate. In the first MMR-based ultrasonic detectors used for photoacoustic imaging the polymer waveguides were directly exposed to the aqueous surrounding [2]. Later, polymer claddings were introduced to protect the waveguide core [3, 4]. It also turned out that by a proper choice of the cladding material, the ultrasonic detection sensitivity can even be increased due to the elasto-optic effect.

Figure 1 illustrates the operation principle of an MRR-based ultrasonic detector with a polymer waveguide core and a polymer cladding layer. The ultrasonic shockwave generated by the absorption of a laser light pulse in tissue at the focal spot induces a change of the resonance wavelength λ_m of MRR. Two physical effects are mainly affecting the resonance condition:

- Local deformations of the waveguide cross section
- Local elasto-optic effect in the waveguide core and in the cladding

We omit the length change of the MRR assuming that the bending of the quartz substrate due to the acoustic pressure wave is negligible. Changes in the coupling strength between the bus and ring waveguide caused by deformations and elasto-optic refractive index changes are also neglected because they impact the resonance condition much less than the effective refractive index changes along the ring. In this study, we focus on the local change of the effective refractive index as a function of acoustic frequency for normal incidence of the acoustic wave. To this end, we employ a 2D FEM multiphysics simulation model that we implemented in COMSOL [5]. The overall resonance shift of the MRR can be estimated by integrating over the full circumference taking into account the local change of the effective refractive index [5, 6]. While previous other studies rely on quasi-static approximations [2, 3, 7], our multiphysics approach takes into account dynamic effects in the mechanical response of the waveguide core and cladding to the acoustic wave as a function of frequency.

We simulate the time dependent deformation of the waveguide structure and the local refractive index change induced by the stress σ due to the elasto-optic effect, which can be described as [8]:

$$n_x \approx n_0 - C_1\sigma_x - C_2(\sigma_y + \sigma_z) \quad (1a)$$

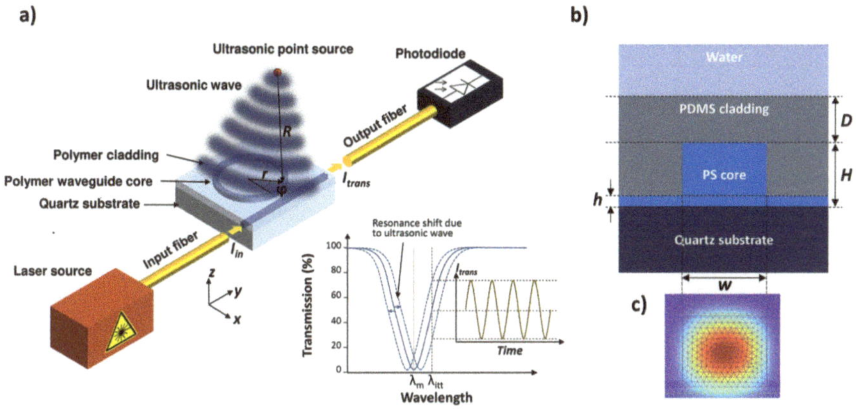

Fig. 1 MRR-based ultrasonic sensor: **a** Operation principle relying on the pressure induced shift of the optical resonance in the MRR interrogated at λ_{itt}, which results in a modulation of the transmitted optical power I_{trans}, **b** cross-section of MRR waveguide ($w = 0.95$ µm, $H = 0.75$ µm, $h = 0.1$ µm), **c** 2D FEM eigenmode analysis of the deformed waveguide structure

$$n_y \approx n_0 - C_1\sigma_y - C_2(\sigma_x + \sigma_z) \tag{1b}$$

$$n_z \approx n_0 - C_1\sigma_z - C_2(\sigma_x + \sigma_y) \tag{1c}$$

In the Eqs. (1a-c), n_x, n_y, and n_z are the refractive indices for light polarized in the x-, y-, and z-direction. n_0 is the refractive index of the material without stress and C_1 and C_2 are the elasto-optic constants. The stress components are denoted by σ, where the subscript defines the stress direction.

Figure 1b *shows the waveguide geometry used for this study with a core width of* $w = 0.95$ µm, height of $H = 0.75$ µm, and residual layer thickness of $h = 0.1$ µm (assuming a nanoimprint fabrication process). Table 1 *summaries the physical constants used in the simulations. The interrogation wavelength* λ_{itt} *was set to 780 nm and the invested mode is the TE-like polarization.* We could not find data for the elasto-optic constants C_1 and C_2 of PDMS in literature. Birefringence measurements for transparent rubbers suggest a relative elasto-optic constant $C_0 = C_2 - C_1$ well beyond 1000 TPa^{-1} [9, 10]. We therefore assumed a difference of 2000 TPa^{-1}.

In our study we use the peak-to-peak change of the effective refractive index n_{eff} of the guided mode during one period of the acoustic wave as indicator for the strength of the MRR response to the acoustic wave. This can be justified by the following relationship. The resonance condition of an MRR is given by

$$m\lambda_m = 2\pi R n_{eff} \tag{2}$$

where R is the radius of the MRR. For a normal incident plane sound wave all parts of the MRR experience the same pressure at the same time. The shift of the resonance

Table 1 Physical constants used in COMSOL multiphysics simulation model (*elasto-optic constants from [11] at 632.8 nm)

Material	Young's modulus E (MPa)	Poisson's ratio ν	Density ρ (kg/m³)	Refractive index n_0	C_1(TPa^{-1})	C_2 (TPa^{-1})
Polystyrene (PS)	300	0.33	965	1.5775	48*	29*
Polydimethylsiloxane (PDMS)	3.0	0.49	1000	1.4050	1000	3000
Glass	7250	0.17	2210	1.4537	–	–

wavelength during a period of the acoustic wave can therefore be expressed as

$$\Delta\lambda = \frac{d\lambda_m}{dP}\Delta P = \frac{\lambda_m^0}{n_{eff}^0}\frac{d\mathrm{n}_{eff}}{dP}\Delta P \tag{3}$$

with ΔP being the pressure amplitude of the sound wave. The parameters λ_m^0 and n_{eff}^0 are the resonance wavelength and the effective refractive index of the undistorted waveguide. The change of the transmitted power I_{trans} can be written as

$$\Delta I_{\mathrm{trans}} = \left.\frac{dI_{\mathrm{trans}}}{d\lambda}\right|_{\lambda_{\mathrm{itt}}} \Delta\lambda. \tag{4}$$

Therefore, the change of effective refractive index per applied pressure $\mathrm{d}n_{\mathrm{eff}}/\mathrm{d}P$ is a direct indicator of the ultrasonic detection sensitivity of the MRR.

5.2 Results and Discussion

The effective refractive index n_{eff} of a guided optical mode is influenced by the whole waveguide layer structure. We therefore investigated the individual contributions of the structural deformation of the waveguide core and of the elasto-optic effects in the core and the cladding layer to the overall change of the effective refractive index. Figure 2 plots n_{eff} over a full period of an acoustic wave with a pressure amplitude of 1 MPa for a set of acoustic frequencies (1, 10, and 100 MHz), where the elasto-optic effect is turned on (EO ON) or off (EO OFF) in the PDMS cladding layer and the waveguide core, respectively. The height D of the cladding layer above the waveguide core is set to 1 μm. Acoustic frequencies of 1, 10, and 100 MHz reveal a similar behavior. We noticed a slight phase delay in the optical response due to the elasto-optic effect at 100 MHz. With respect to the effective refractive index of the undeformed waveguide n_{eff}^0 (baseline indicated by a horizontal line in Fig. 2) compressive strain reduces n_{eff}, ignoring the elasto-optic effect in the cladding

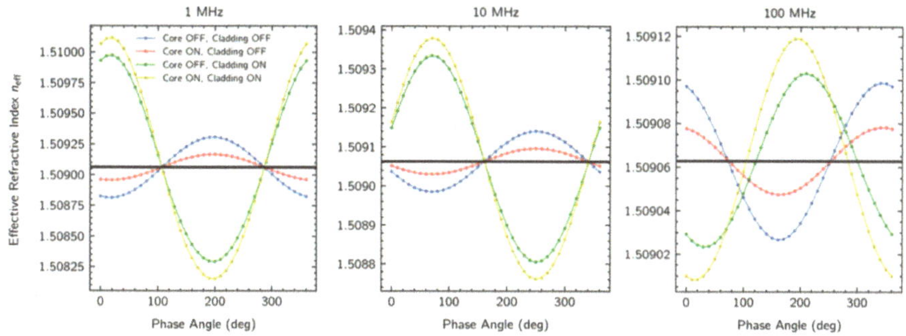

Fig. 2 Contribution of the elasto-optic effect in the PS waveguide core and the PDMS cladding to the change of n_{eff} over a full period of an acoustic wave with 1 MPa pressure amplitude for a cladding thickness of 1 μm at ultrasonic frequencies from 1, 10, and 100 MHz. The black horizontal lines indicate the baseline of the effective refractive index for the undeformed waveguide n_{eff}^{0}. The elasto-optic effect of the cladding layer (Cladding ON) dominates the optical response

layer and waveguide core, because the core becomes smaller and the mode expands stronger into the low refractive index cladding.

The elasto-optic effect in the waveguide core increases n_{eff}, although the geometry effect still dominates the overall change of n_{eff} if the elasto-optic contribution of the polymer cladding is omitted. However, the elasto-optic effect in the cladding layer has the strongest influence on the overall change of n_{eff}.

Next, we investigated the frequency dependence of the peak-to-peak change of n_{eff} caused by ultrasonic sound waves with 1 MPa pressure amplitude for different cladding heights ($D = 1, 3, 5, 10$ μm) in the frequency range from 0.1 to 100 MHz. Figure 3 plots the results. At low frequencies of 0.1 MHz all structures behave in the same way. This is the frequency region where the quasi-static model still applies for all cladding thicknesses. For small cladding thicknesses of 1 to 3 μm the peak-to-peak change of n_{eff} stays constant up to a frequency of 1 MHz. With further increasing frequency and especially in the range from 1 to 10 MHz the peak-to-peak change of n_{eff} decreases constantly. Moreover, a thin cladding layer gives a stronger response. In the frequency region above ~6 MHz we notice the occurrence of peaks at certain frequencies that depend on the height of the cladding layer. These peaks are related to standing waves that form in the cladding layer.

To further underpin this hypothesis, we investigated the displacement of the cladding material within a cladding of 10 μm thickness at the first (~10 MHz) and second (~22 MHz) resonance peak, and between (17 MHz). Figure 4 shows the waveguide systems at an acoustic phase angle of 0°. The compression of the waveguide cladding is more pronounced at the resonance peaks. The results of the frequency sweep and material displacement show that the formation of standing acoustic waves in the cladding can have a major influence on the sensitivity of the acoustic waveguide sensor.

The acoustic resonance effect observed in these simulations is directly related to the elastic properties of the polymer cladding layer. In particular, Poisson's ratio plays

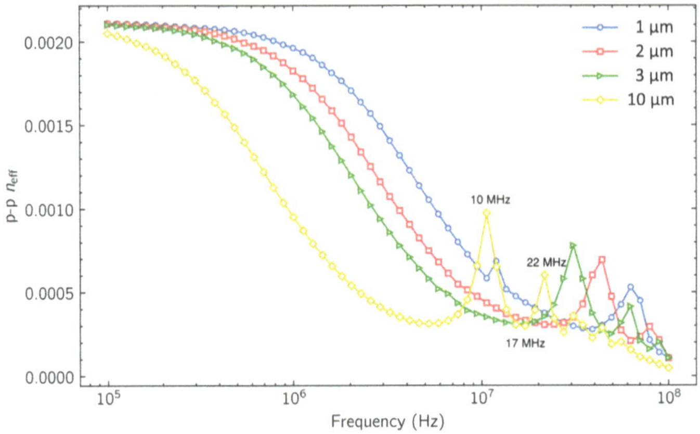

Fig. 3 Frequency dependent peak-to-peak change of n_{eff} over a full period of an acoustic wave with 1 MPa pressure amplitude for various cladding thicknesses (1, 2, 3, 10 µm). For different cladding thicknesses acoustic resonance peaks occur at different frequencies

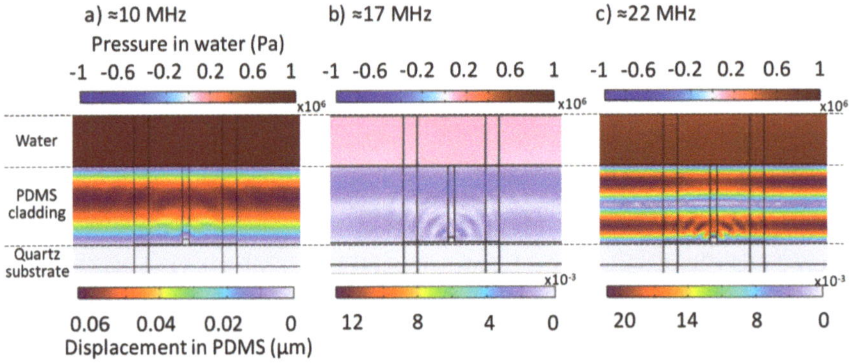

Fig. 4 Displacement of material within the PDMS cladding with 10 µm thickness at an acoustic phase angle of 0° (*i.e.*, max. Pressure of 1 MPa in water): **a** first resonance peak at ~10 MHz, **b** between the resonance peaks at 17 MHz, **c** second resonance peak at ~22 MHz. The color bars on the top and bottom show the acoustic pressure in water and the displacement of material within the cladding, respectively

a critical role in the formation of these standing waves in the cladding layer, which becomes obvious from the equations for the acoustic wavelength in an isotropic medium

$$\lambda_{\mathrm{US}} = \frac{c_l}{f} = \frac{1}{f}\sqrt{\frac{E}{\rho}\frac{(1-\nu)}{(1+\nu)(1-2\nu)}} \tag{5}$$

where the speed of longitudinal sound waves is given by

$$c_l = \sqrt{\frac{E}{\rho} \frac{(1-\nu)}{(1+\nu)(1-2\nu)}} \tag{6}$$

with E being Young's modulus, ρ the density, and ν the Poisson's ratio of the cladding material. For the assumed material parameters of PDMS (see Table 1) the wavelength of longitudinal sound waves amounts to $\lambda_{US} \sim 20$ µm at a frequency of 10 MHz. The fundamental standing wave between two hard boundaries (*i.e.*, incompressible materials) forms for a cavity length of $\lambda_{US}/2$, which corresponds to the case depicted in Fig. 3a. Since the Poisson's ratio of PDMS is close 0.5, small changes have significant impact on the sound velocity and, thus, on the acoustic resonance frequencies in the cladding. Additional simulations indicated that as the Poisson's ratio approaches 0.5 the ultrasonic detection bandwidth expands to higher frequencies. At the same time, according to Eq. (5) standing waves can only occur at much thicker claddings.

5.3 Conclusion and Outlook

We investigated the time dependent optical response of a polymer waveguide structure of an MRR in terms of effective refractive index change of the guided mode to an incident acoustic plane wave using a 2D multiphysics FEM simulation model. This allowed us to gain insight into the interdependencies of the optical response of an MRR to a normally incident acoustic wave with the cladding thickness and the acoustic frequency. The results indicate that the elastic properties of the cladding material, in particular the Poisson's ratio, play a key role in the frequency response. The elastic properties of PDMS strongly depend on the processing parameters [12, 13]. Therefore, precise control over fabrication is essential. Simulation results also suggest that a high Poisson's ratio close to 0.5 is advantageous in terms of ultrasonic detection bandwidth. Further simulations will be carried out to substantiate this hypothesis.

Moreover, since the elasto-optic effect of the PDMS cladding significantly contributes to the overall optical response of the MRR, which has also been investigated experimentally by Erdogan et al. [7], and the fact that the pure structural deformation of the waveguide core has an opposite effect on the effective index change of the guided mode suggests the use of a stiff core material, which not necessarily has to be a polymer. Inorganic core materials also offer the advantage of a higher index contrast and thus narrower ring radii can be realized, which is beneficial for a larger acceptance angles of the incident acoustic wave (field of view) at high acoustic frequencies [6].

Acknowledgements This work was supported by the REAP project, which received funding from the European Union's Horizon 2020 research and innovation programme under grant agreement No 101016964 under the Photonics Public Private Partnership (PPP) call H2020-ICT-2020-2 with the topic ICT-36-2020—Disruptive photonics technologies.

References

1. Lee, Y., Zhang, H.F., Sun, C.: Highly sensitive ultrasound detection using nanofabricated polymer micro-ring resonators. Nano Converg. **10**(1), 30 (2023)
2. Chao, C.-Y., Ashkenazi, S., Huang, S-W., O'Donnell, M., Guo, L. J.: High-frequency ultrasound sensors using polymer microring resonators. IEEE Trans. Ultrason. Ferroelectr. Freq. Control 54(5), 957–965 (2007)
3. Zhang, S., Chen, J., He, S.: Novel ultrasound detector based on small slot micro-ring resonator with ultrahigh Q factor. Opt. Commun. **382**, 113–118 (2017)
4. Li, H., Dong, B., Zhang, X., Shu, X., Chen, X., Hai, R., Czaplewski, D.A., Zhang, H.F., Sun, C.: Disposable ultrasound-sensing chronic cranial window by soft nanoimprinting lithography. Nat. Commun. **10**(1), 4277 (2019)
5. Schumacher, J.-J., Muellner, P., Eggeling, M., Hainberger, R.: Multiphysics simulation approach for polymer waveguide microring resonator based ultrasonic detectors. In: In: 24th European Conference on Integrated Optics, ECIO 2023, ID 192 (2023)
6. Zhang, Z., Dong, B., Li, H., Zhou, F., Zhang, H.F., Sun, C.: Theoretical and experimental studies of distance dependent response of micro-ring resonator-based ultrasonic detectors for photoacoustic microscopy. J. Appl. Phys. **116**(14), 144501 (2014)
7. Erdogan, R.T., Filonenko, G.A., Picken, S.J., Steeneken, P.G., Westerveld, W.J.: Optimizing cladding elasticity to enhance sensitivity in silicon photonic ultrasound sensors. J. Lightwave Technol. **43**(3), 1419–1428 (2025)
8. MacAdam, D.L., Theocaris, P.S., Gdoutos, E.E. (eds.): Matrix Theory of Photoelasticity. Springer, Berlin Heidelberg (1979)
9. Angioletti, A., Eccher, S., Polvara, O., Zerbini, V.: Rubber birefringence and Photoelasticity. Rubber Chem. Technol. **38**(5), 1115–1163 (1965)
10. Hazan, Y., Levi, A., Nagli, M., Rosenthal, A.: Silicon-photonics acoustic detector for optoacoustic micro-tomography. Nat. Commun. **13**(1), 1488 (2022)
11. Ay, F., Kocabas, A., Kocabas, C., Aydinli, A., Agan, S.: Prism coupling technique investigation of elasto-optical properties of thin polymer films. J. Appl. Phys. **96**(12), 7147–7153 (2004)
12. Müller, A., Wapler, M.C., Wallrabe, U.: A quick and accurate method to determine the Poisson's ratio and the coefficient of thermal expansion of PDMS. Soft Matter. **15**(4), 779–784 (2019)
13. Smith, A.M., Inocencio, D.G., Pardi, B.M., Gopinath, A., Andresen Eguiluz, R.C.: Facile determination of the Poisson's ratio and Young's modulus of polyacrylamide gels and polydimethylsiloxane. ACS Appl. Polym. Mater. **6**(4), 2405–2416 (2024)

Chapter 6
All-Silicon Cascaded Microring Resonator Photodetector for 160 Gbps Operation in the C-Band

Zelu Wang and Hon Ki Tsang

Abstract We propose and demonstrate an all-silicon cascaded microring resonator photodetector operating in the C-band. By harnessing cavity enhancement with avalanche multiplication, the device achieves a responsivity of 0.38 A/W at − 8.4 V bias. The dark current is around 300 nA. The cascaded microring structure significantly broadens the device bandwidth, enabling high-speed data transmission at 100 Gbit/s NRZ and 160 Gbit/s PAM4 under the general SD-FEC threshold. This performance marks the highest data rate reported to date for C-band all-silicon photodetectors.

6.1 Introduction

Silicon photonics has emerged as a critical technology in applications such as high-speed optical interconnects, computing, and sensor systems, owing to its high-density integration and CMOS compatibility. However, bulk silicon is inherently transparent in the near-infrared region because its photon energy lies below the material's 1.1 eV bandgap (corresponding to a wavelength of ~ 1.1 μm). While this optical transparency benefits light guiding, it poses a significant challenge for light detection. Consequently, integrated photodetectors for optical-to-electronic conversion have typically relied on germanium or hybrid integration of III-V semiconductors, adding additional processes and costs to the chip fabrication [1].

Therefore, there have been research efforts on developing silicon-based sub-bandgap photodetectors with simpler fabrication processes. Although such silicon waveguide photodetectors offer a fabrication advantage, their lower responsivity and absorption coefficients necessitate longer device lengths [2]. All-silicon sub-bandgap photodetectors based on the combined effects of two-photon absorption, surface state absorption, and photon-assisted tunneling have been previously studied

Z. Wang · H. K. Tsang (✉)
Department of Electronic Engineering, The Chinese University of Hong Kong, Shatin, New Territories, Hong Kong SAR, China
e-mail: hktsang@ee.cuhk.edu.hk

P. Smowton and T. Krauss (eds.), *Proceedings of the 26th European Conference on Integrated Optics*, Springer Proceedings in Physics 440,
https://doi.org/10.1007/978-3-032-29673-3_6

for O-band [3, 4]. Optical resonant enhancement can enable such photodetectors to attain responsivity comparable to that of extended straight waveguide detectors, thereby improving overall performance. Extending the all-silicon photodetectors for operation to the C-band is more challenging due to the longer wavelengths being further from silicon's bandgap energy, resulting in reduced absorption efficiency from the photon assisted tunneling process. To date, the reported C-band devices have been limited to data rates of 40 Gb/s and responsivities around 0.05 A/W [5].

In this work, we present the first experimental demonstration of a high-speed, cascaded microring resonator photodetector (CMRR-PD) operating in the C-band. The device, structured with two microrings in tandem, not only enhances bandwidth but also maintains a compact footprint of 30 × 60 μm^2. Fabricated using CMOS-compatible processes in a commercial foundry via a standard multi-project wafer shuttle run, the device achieves a responsivity of 0.38 A/W at − 8.4 V bias with a dark current of 300 nA. Furthermore, we demonstrate the data transmission performance at 100 Gbit/s NRZ and 160 Gbit/s PAM4 under the general SD-FEC threshold. Additionally, the resonant nature allows for inherent wavelength filtering, rendering it highly suitable for wavelength division multiplexing systems and next-generation optical communication networks.

6.2 Device Design and Analysis

Figure 6.1a displays a microscope image of the cascaded microring resonator-based photodetector featuring a single-side coupled bus waveguide. The two identical rings, each with 8-μm bending radius, were implemented with bent directional couplers to enable broadband operation. The PN junction is embedded within a 500 nm-wide rib waveguide with a 90 nm-thick slab, and heavily doped regions are positioned 750 nm from each waveguide edge to form ohmic contacts with the aluminum electrodes. Two integrated heaters facilitate precise resonance tuning. For sub-bandgap light detection, photocurrent is generated via a combination of two-photon absorption, surface state absorption, and photon-assisted tunneling in the silicon PN junction under high reverse bias [6]. Figure 6.1c illustrates the simulated electric field distribution at − 8.4 V, confirming that the field strength is sufficient to trigger photon-assisted tunneling and amplify the photocurrent through avalanche multiplication.

The microring cavity enhances the internal optical field at resonance wavelengths, thereby boosting photocurrent generation. In a high-Q microring resonator photodetector, this cavity enhancement yields high responsivity; however, the associated long photon lifetime limits the device's OE bandwidth. By contrast, our cascaded microring resonator photodetector employs moderate inter-ring detuning, which mitigates avalanche gain saturation and allows it to achieve responsivity comparable to that of a single-ring device at resonance, while significantly improving the OE bandwidth. This cascaded configuration effectively reduces the photon lifetime without sacrificing the benefits of cavity enhancement. Furthermore, the small-signal response of the CMRR PD, derived using time-domain coupled mode theory

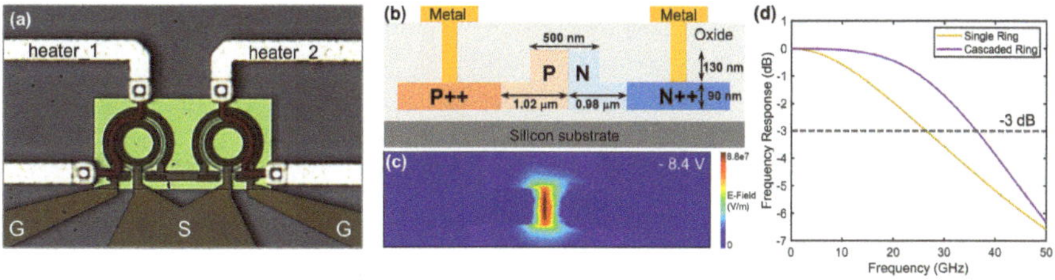

Fig. 6.1 **a** Microscope image of the CMRR photodetector; **b** cross-sectional schematic of the ring waveguide; **c** simulated electric field distribution in the rib waveguide under − 8.4 V bias voltage; **d** simulated frequency response of single ring and cascaded ring models

[7], shown in Fig. 6.1d, confirming that the cascaded architecture offers a substantial bandwidth improvement over the single-ring model while maintaining similar responsivity.

6.3 Experimental Results

The chip-to-fiber coupling was implemented using grating couplers, which introduce an insertion loss of approximately 6 dB per fiber interface. Figure 6.2a presents the measured transmission spectrum at − 8.4 V bias, where fine-tuning of the integrated heaters enabled a split resonance with an inter-ring detuning of 170 pm. Although measurements were conducted at a single wavelength of 1538.2 nm, the tunability of the resonances via integrated heaters allows coverage of the entire free spectral range (FSR) of 12.15 nm, demonstrating its potential as a demultiplexing photodetector in wavelength-division multiplexing (WDM) systems. The photocurrent response of the CMRR photodetector was characterized across the entire spectrum under varying reverse bias voltages from 0 V to − 8.4 V, as depicted in Fig. 6.2b. A significant increase in photocurrent was observed around the two resonance wavelengths. The dark current, shown in Fig. 6.2c, remained relatively low, ranging from 15 nA at − 8 V to 300 nA at − 8.4 V, which is substantially lower than that of conventional germanium photodetectors. The incident optical power on the photodetector was approximately − 15 dBm. The device responsivity, also illustrated in Fig. 6.2c, exhibited a rapid increase as the reverse bias voltage increased from − 7 V to − 8 V due to avalanche multiplication, reaching a peak value of 0.38 A/W at − 8.4 V.

Data transmission tests were performed using the experimental setup depicted in Fig. 6.3a. A 13 dBm external cavity laser operating at 1538.2 nm served as the optical source. A 256 GSa/s arbitrary waveform generator (Keysight M8199A) produced a PRBS-15 electrical signal, which was then amplified by a 65 GHz linear RF amplifier before being applied to a 40 GHz $LiNbO_3$ modulator. To compensate for fiber-to-chip coupling loss, an erbium-doped fiber amplifier (EDFA) was employed before coupling the optical signal into the device under test (DUT). The CMRR photodetector was electrically probed using a 65 GHz RF probe, with a bias-T supplying

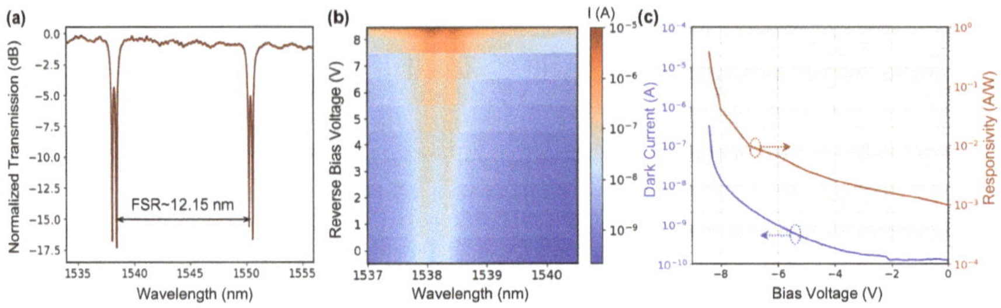

Fig. 6.2 **a** Measured transmission spectrum of the cascaded microring resonator with 170 pm inter-ring detuning under − 8.4 V bias voltage; **b** measured 2D colormap of photocurrent versus reverse bias voltage and wavelength; **c** dark current and responsivity as functions of bias voltage

the DC bias and extracting the RF signal. The detected electrical signal was then captured by a 59 GHz, 256 GSa/s real-time oscilloscope (Keysight UXR0592). As shown in Fig. 6.3b, the bit error rate (BER) performance of the 100 Gbit/s NRZ signal was evaluated under varying optical power levels, demonstrating a sensitivity of − 2 dBm under the SD-FEC threshold. The measured eye diagrams and count distributions for 100 Gbit/s NRZ and 160 Gbit/s PAM4 signals (bandwidth limited by the 40 GHz modulator) at − 8.4 V bias voltage are presented in Fig. 3c and d, respectively. A bandpass filter can be incorporated to suppress amplified spontaneous emission noise induced by EDFA, thereby enabling higher data rates. Additionally, OE bandwidth characterization of the device will be conducted in the future.

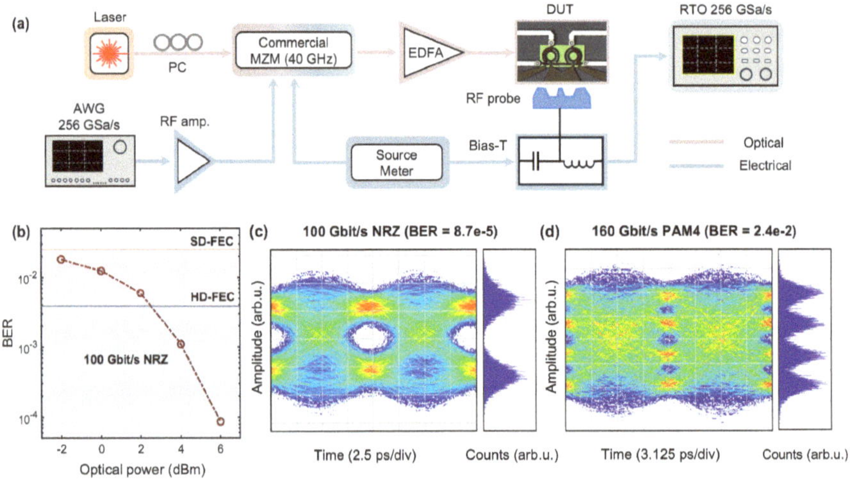

Fig. 6.3 **a** Experimental setup for the data transmission test; **b** measured bit error rate (BER) versus input optical power of 100 Gbit/s NRZ signal; Measured eye diagrams and count distribution of **c** 100 Gbit/s NRZ signal and **d** 160 Gbit/s PAM4 signal at bias voltage of − 8.4 V

6.4 Conclusion

In conclusion, we have demonstrated an all-silicon cascaded microring resonator photodetector operating in the C-band that achieves a responsivity of 0.38 A/W at −8.4 V bias and a dark current of 300 nA. The cascaded microring architecture substantially enhances the device bandwidth, facilitating high-speed data transmission at 100 Gbit/s NRZ and 160 Gbit/s PAM4, which represents the highest data rate reported for C-band all-silicon photodetectors. Additionally, the device features a compact footprint of $30 \times 60\ \mu m^2$ and is fabricated using CMOS-compatible processes. The device exhibits considerable potential for integration into next-generation silicon photonic integrated circuits and advanced optical communication systems.

Acknowledgements This work was supported by RGC General Research Fund Grant 14204623. Zelu Wang also thanks RGC for funding support from the Hong Kong PhD Fellowship scheme.

References

1. Shi, Y., Li, X., Chen, G., Zou, M., Cai, H., Yu, Y., Zhang, X.: Avalanche photodiode with ultrahigh gain–bandwidth product of 1,033 GHz. Nat. Photonics. **18**(6), 610–616 (2024)
2. Grote, R.R., Padmaraju, K., Souhan, B., Driscoll, J.B., Bergman, K., Osgood, R.M.: 10 Gb/s error-free operation of all-silicon ion-implanted-waveguide photodiodes at 1.55 μm. IEEE Photon. Technol. Lett. **25**(1), 67–70 (2012)
3. Peng, Y., Yuan, Y., Sorin, W.V., Cheung, S., Huang, Z., Hong, C., Beausoleil, R.G.: An 8× 160 Gbs^{-1} all-silicon avalanche photodiode chip. Nat. Photonics. **18**(9), 928–934 (2024)
4. Sakib, M., Liao, P., Kumar, R., Huang, D., Su, G.L., Ma, C., Rong, H.: A 112 Gb/s all-silicon micro-ring photodetector for datacom applications. In: Optical Fiber Communication Conference, p. Th4A-2. Optica Publishing Group (2020)
5. Li, X., Li, Z., Xiao, X., Xu, H., Yu, J., Yu, Y.: 40 Gb/s all-silicon photodetector based on microring resonators. IEEE Photon. Technol. Lett. **27**(7), 729–732 (2015)
6. Casalino, M., Coppola, G., De La Rue, R.M., Logan, D.F.: State-of-the-art all-silicon sub-bandgap photodetectors at telecom and datacom wavelengths. Laser Photonics Rev. **10**(6), 895–921 (2016)
7. Gu, T., Chen, Y.K., Wong, C.W., Dong, P.: Cascaded uncoupled dual-ring modulator. Opt. Lett. **39**(16), 4974–4977 (2014)

Chapter 7
Bandwidth and Power Conversion Efficiency Analysis of Kerr Frequency Combs in Silicon Nitride Microresonators

M. W. Balkie, A. Nesic, and A. L. Schall-Giesecke

Abstract Kerr frequency combs generated in silicon nitride microring resonators contribute to advancements in integrated photonics. Understanding the performance efficiency and the operating behavior of frequency combs is essential for the systematic design of microring resonators. In this paper we investigate the pump power and dispersion effects on the spectral bandwidth (BW) and on the power conversion efficiency (PCE) of silicon nitride microresonators based on a modified coupled mode equation to include thermal effects. The formation of a frequency comb in different operation regimes, i.e., modulation instability (MI), multi-soliton (MS), and single soliton states, is numerically studied for a resonator under different input pump power, input pump frequency to ring resonance detuning, and different anomalous dispersions. The PCE and BW analysis results show that an increase in pump power enhances the BW, while the PCE decreases. On the other hand, the increase in second order dispersion decreases the BW, while the PCE improves. The results therefore, indicate that there is a trade-off between the PCE and the BW of single-solitons. Our analysis shows the critical importance of the selection of waveguide dimensions and operational parameters to optimize the performance of Kerr frequency combs.

M. W. Balkie (✉) · A. Nesic · A. L. Schall-Giesecke
Fraunhofer Institute for Microelectronic Circuits and Systems (IMS), Duisburg, Germany
e-mail: mulualem.wale.balkie@ims.fraunhofer.de

A. L. Schall-Giesecke
Department of Electrical Engineering and Information Technology, University of Duisburg-Essen (UDE), Duisburg, Germany

Center for Nanointegration Duisburg-Essen (CENIDE), Duisburg, Germany

P. Smowton and T. Krauss (eds.), *Proceedings of the 26th European Conference on Integrated Optics*, Springer Proceedings in Physics 440,
https://doi.org/10.1007/978-3-032-29673-3_7

7.1 Introduction

Kerr frequency combs, generated in high Q-factor microresonators, have emerged as powerful tools for different applications, including optical communications [1], quantum optics [2], precision metrology [3], spectroscopy [4], and optical clocks [5]. The generation of stable Kerr frequency combs relies on the interplay between dispersion and nonlinearity within the resonator. In particular, operating within the anomalous dispersion regime allows for the generation of dissipative Kerr soliton (DKS) microcombs with coherent optical comb lines [6–8]. Attaining high power conversion efficiency (PCE) and bandwidth (BW) is a critical challenge on Kerr combs [9, 10]. The maximum attainable BW for the temporal cavity solitons was analytically estimated using the solutions of the Lugiato-Lefever equation [11]. Moreover, the spectral efficiencies dynamical properties were investigated for a dual-tone-pumping using the Lugiato-Lefever equation [12]. The power conversion efficiency of the Kerr comb has been studied numerically using modified coupled mode equation to include thermal effects [13]. However, a comprehensive study of the PCE and the BW for different dispersion and pump power is important. In this paper, we investigate the dependencies of the PCE and the BW of Kerr combs on different pump power and dispersion values in a generic silicon nitride (SiN) platform in the anomalous dispersion regime.

7.2 Theoretical Model Description

The resonance frequencies of the microresonator (ω_μ) can be written as the Taylor expansion around the pump mode ω_0 [6]:

$$\omega_\mu = \omega_0 + D_1\mu + \frac{1}{2}D_2\mu^2 + \frac{1}{6}D_3\mu^3 + \ldots \tag{7.1}$$

where μ is the mode number referenced to the center mode number ($\mu = 0$ corresponds to the pump mode), $D_1/2\pi$ is the free spectral range (FSR) of the microresonator and $D_2/2\pi$ is the second order dispersion, related to the group velocity dispersion (GVD), $D(\lambda) = -2\pi c\beta_2(\omega)/\lambda^2$, where $\beta_2 = -n/cD_2/D_1^2$; n and c are the refractive index and the speed of light, respectively. For anomalous GVD ($D(\lambda) > 0$), third order Kerr nonlinearity leads to a nonlinear coupling between different resonance modes, which results in spectral broadening through self-phase modulation [6]. To achieve anomalous GVD, SiN waveguide dimensions are determined by simulations using finite difference eigenmode (FDE) solver—the results of the simulations are presented in Fig. 7.1. Figure 7.1a shows a normalize mode field distribution of the fundamental TE optical mode at a wavelength of 1550 nm. Figure 7.1b shows GVD values for different waveguide width w, and height h—in general anomalous GVD is achieved with thicker SiN layers. Point A, corresponds

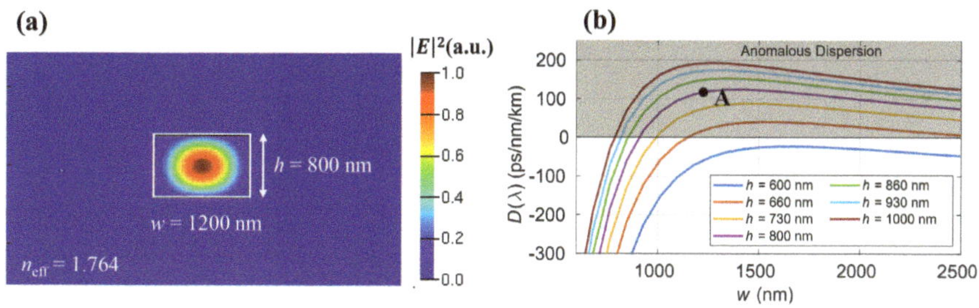

Fig. 7.1 **a** Normalized mode field distribution of the fundamental transverse electric (TE) mode (TE0) in a SiN waveguide (core silicon and cladding silicon dioxide at a wavelength of 1550 nm). **b** Simulated group velocity dispersion (GVD), D (λ) (ps/nm/km) of a SiN ring resonator with a radius of 114 μm, for different waveguide dimensions at 1550 nm wavelength

to $w = 1200$ nm and $h = 800$ nm, which are the dimensions of the waveguide shown on Fig. 7.1a.

The intracavity field evolution in the microresonator and the corresponding Kerr frequency comb generation can be described using a coupled mode equation written in a dimensionless way [7, 14], see Eq. (7.2). For considering thermal effects a normalized temperature variation equation is added for the numerical simulation [13], see Eq. (7.3).

$$\frac{\partial a_\mu}{\partial \tau} = -\left(1 + j\left(\zeta_0 + d_2\mu^2\right) - j\Theta\right)a_\mu + j\sum_{\mu' \le \mu''}\left(2 - \delta_{\mu'\mu''}\right)a_{\mu'}a_{\mu''}a^*_{\mu'+\mu''-\mu} + \delta_{0\mu}f \tag{7.2}$$

$$\frac{\partial \Theta}{\partial \tau} = \frac{2}{\kappa \tau_{\mathrm{T}}}\left(\frac{n_{2\mathrm{T}}}{n_2}\sum_{\mu}\left|a_\mu\right|^2 - \Theta\right) \tag{7.3}$$

In Eq. (7.2), a_μ is the normalized slowly varying amplitude of the comb modes close to the mode frequency ω_μ, τ is the normalized time, ζ_0 is the normalized detuning, d_2 is the normalized second-order dispersion, μ is the relative mode number with respect to the pumped mode, f is the normalized pump amplitude, and $\delta_{0\mu}$ is the Kronecker delta (see Appendix A for the relations between the normalized and physical parameters). In Eq. (7.3), Θ is the normalized variation of temperature. τ_{T} is the thermal relaxation time and $n_{2\mathrm{T}}$ is the coefficient of thermal nonlinearity.

PCE in Kerr combs quantify how effectively the input pump power is converted into the comb lines. It is defined as [15, 16]:

$$\mathrm{PCE} = \frac{P_{\mathrm{comb}}}{P_{\mathrm{pump}}} \tag{7.4}$$

where P_{comb} is the power of the generated comb lines excluding the center mode which is from the pump and P_{pump} is the input pump power. The spectral bandwidth of the comb is defined as the frequency range in which the power remains within 3 dB of the power $|a_1|^2$ of the mode directly adjacent to the pump mode [11]:

$$\mathrm{BW} = f_{\max} - f_{\min} \tag{7.5}$$

In Eq. (7.5), $f_{\max}$ is the upper frequency bound and $f_{\min}$ is the lower frequency bound at which the comb power drops by 3 dB from the power of the mode directly adjacent to the pump mode. Our simulation builds on the open-source solver from [17], with added scripts for BW and PCE analysis for the Kerr combs. The simulation parameters used are listed in the Appendix A. In this simulation higher order dispersions are not considered.

7.3 Intracavity Field Dynamics and Spectral Efficiency Analysis

The two main controlling parameters for the intracavity field evolution are the normalized pump field amplitude f and the detuning ζ_0.. The latter can be done with tuning the pump laser relative to the cavity resonance. When the pump laser is swept from the higher frequency to the lower frequency of the cavity resonance (called forward tuning), the intracavity field results in a triangular trace [14]. This approach can lead to formation of multiple intracavity solitons when the pump enters the effective red-detuned regime, i.e., the pump laser tuned through the effective zero detuning point [7]. The simulation result of forward tuning of the laser pump is as shown in Fig. 7.2 for pump power $P_{\mathrm{pump}} = 2$ W and $D_2 = 1.4$ MHz. The evolution of average intracavity intensity as a function of forward laser tuning shows different dynamical states in the microresonator, see Fig. 7.2a. It starts in the continuous-wave (CW) state followed by the modulation instability (MI) regime, and chaotic state and multi-soliton (MS) state. The intracavity power over one complete round trip and the corresponding optical spectrum are illustrated for the CW state Fig. 7.2b and c, MI comb state Fig. 7.2d and e, chaotic comb state Fig. 7.2f and g, and multi-soliton state Fig. 7.2h and i. In the MS regime the position of solitons remains stable and the number of solitons can be reduced with further tuning [7].

Reaching to a single soliton state in only forward tuning is rarely possible and a backward tuning technique (tuning from lower frequency to higher frequency) is used with a relatively small tuning steps after the MS state is reached with forward tuning [13, 14]. In backward tuning, the number of solitons is successively reduced to the single-soliton state, see Fig. 7.3a, which has a well-defined spectral property. The number of soliton reduction steps depends on the tuning speed—a slow tuning speed results in the removal of one soliton at a time, whereas a faster tuning speed may cause some solitons to be skipped. The reduction of number of solitons at different steps and their corresponding optical spectrum is illustrated in Fig. 7.3b to i.

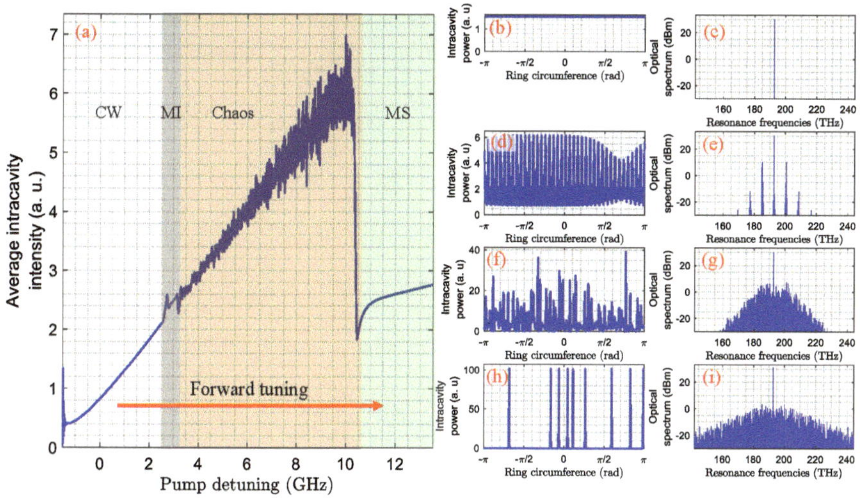

Fig. 7.2 Intracavity field dynamics during forward laser tuning, with simulation parameters listed in the main text and in Appendix A. **a** Evolution of average intracavity intensity as a function of pump detuning. It starts in the CW state with a constant intracavity power shown in **b** and a single frequency as shown in **c**. The light gray shaded regions indicate the modulation instability state where sidebands start to generate and the intracavity power and the corresponding optical spectrum are shown in **d** and **e**. The light red shaded region indicates the chaotic state and the intracavity power, and its optical spectrum are shown in **f** and **g**. The light green shaded regions indicate the formation of the multi-solitons, the intracavity power and the corresponding optical spectrum are shown in **h** and **i**

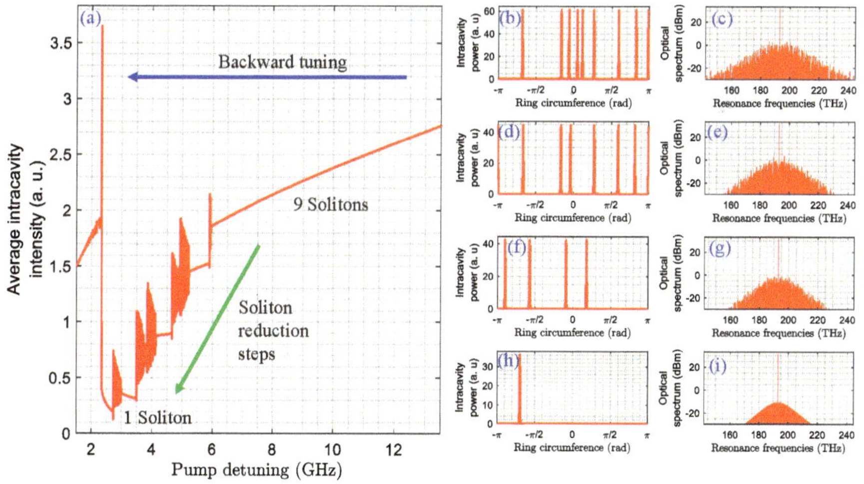

Fig. 7.3 Intracavity field dynamics during backward laser tuning, with simulation parameters listed in the main text and in Appendix A. **a** Average intracavity intensity as a function of backward laser detuning. The system transitions from a multi-soliton state to a single soliton state through a series of soliton reduction steps. The intracavity power and the corresponding optical spectrum are illustrated for: **b** and **c** nine solitons, **d** and **e** seven solitons, **f** and **g** four solitons, and **h** and **i** a single soliton

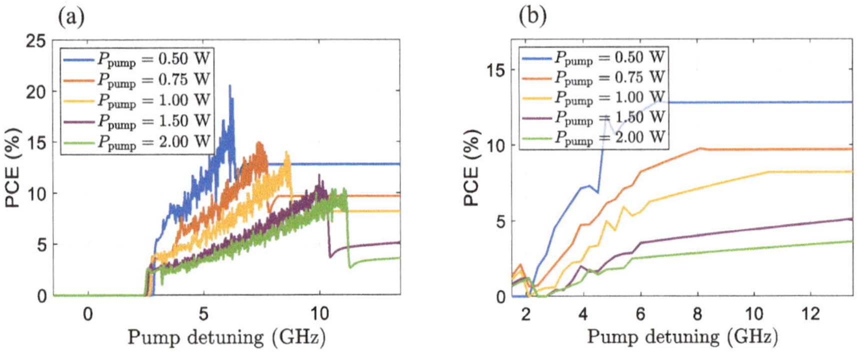

Fig. 7.4 PCE for forward and backward tuning. **a** PCE comparisons for different pump powers for the combs generated in forward tuning. PCE is zero until the tuning reaches the MI state and reaches as high as 20% in the chaotic state of the comb for 0.5 W pump power. **b** PCE comparisons for soliton combs in the backward tuning. The PCE for the multi-soliton state (nine-solitons) reaches 13% for a 0.5 W pump power and it decreases for higher pump powers reaching 3% for 2 W. When the number of solitons reduces, the PCE also reduces and it drops less than 1% for a single soliton state

Spectral efficiency analysis. The spectral efficiency of Kerr combs is key for different applications. The PCE of the comb is calculated using Eq. (7.4) and the results are presented in Fig. 7.4 as a function of detuning at $D_2 = 1.4$MHz for different pump power values. The PCE remains at zero until the laser detunes into the modulation instability (MI) region, after which it increases in the chaotic comb regime. However, upon reaching MS state, the PCE decreases. In addition, increasing the pump power results in reduction in PCE, decreasing from 13% at 0.5 W to 3% at 2 W pump power in the multi-soliton state. Figure 7.4b illustrates the PCE of the combs in backward tuning of the pump, where PCE decreases when the number of solitons decrease and reaches below 1% for a single soliton state due to the gradual spectral compaction of the comb.

For the single soliton state, we use Eq. 7.4 to calculate the PCE and Eq. 7.5 to calculate the BW, as functions of pump power and second order dispersion—the results are presented in Fig. 7.5. In a first step, we increase the pump power P_{pump} from 0.75 W to 2.25 W, while keeping the dispersion D_2 constant at $D_2 = 1.4$MHz. The PCE decreases from 0.7% to 0.3%, while the BW increases from 9 to 16 THz, see Fig. 7.5a. In a second step, we keep the pump power P_{pump} constant at $P_{pump} =$ 2 W, and increase the dispersion D_2 from 1.25 MHz to 2.3 MHz. The PCE increases from 0.28% to 0.45%, while the BW decreases from 15.5THz to 11.25 THz, see Fig. 7.5b. These results show that there is a trade-off between the PCE and the BW, and this must be carefully considered when designing microresonators for Kerr comb generation.

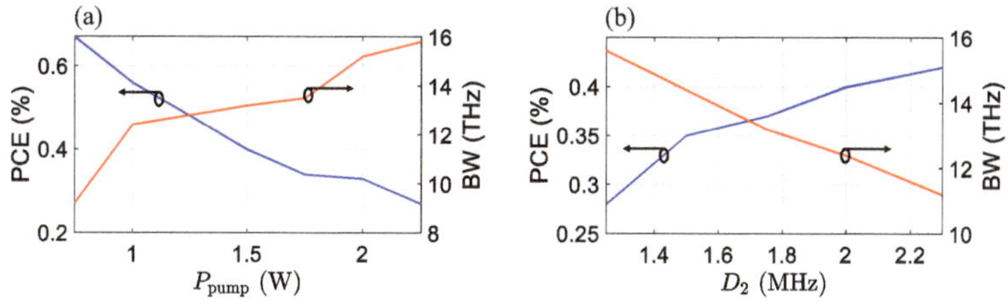

Fig. 7.5 PCE and BW for a single soliton comb. **a** PCE (left axis) and BW (right axis) as functions of P_{pump}. The PCE decreases, and the BW increases when the P_{pump} increases. **b** PCE (left axis) and BW (right axis) as functions of D_2. The PCE increases, and the BW decreases when D_2 increases

7.4 Conclusion

We present numerical simulation results of the key performance metrics, the power conversion efficiency (PCE) and bandwidth (BW), of Kerr frequency combs in silicon nitride microring resonators within the anomalous dispersion regime using a modified coupled mode equation to include thermal effects. Different conditions of pump power (P_{pump}), detuning, and second-order dispersion (D_2) have been studied in different states of Kerr combs, i.e., MI-state, MS-state and single soliton state. In our numerical experiment, we find that PCE decreases with increasing P_{pump} and increases with increasing D_2. On the other hand, we find that the BW of the single soliton combs increases with increasing P_{pump} and decreases when D_2 increases. These results indicate a trade-off between the PCE and the BW, indicating the importance of optimizing waveguide dimensions and operational parameters (P_{pump} and detuning) to achieve desired performance in microresonator based frequency combs.

Acknowledgements This work was supported by "NANO-INC". The project "Nanolaminate für innovative nichtlineare nanophotonische Schaltkreise (NANO-INC, FKZ 13N17104)" is funded by the Federal Ministry of Education and Research of Germany.

This work was supported by the project "FREDERIC". The project "FouRiEr Domain ElectRIcal-to-optical Converter "funded by German Research Foundation (DFG)—Project number 471599135.

Appendix A

In Eq. (2), $a_\mu = A_\mu\sqrt{2g/\kappa}$ where A_μ are the comb mode amplitudes (μ is an integer), κ is the cavity decay rate and g is the nonlinear coupling coefficient

$$g = \frac{\hbar\omega_0^2 c n_2}{n_0^2 V_{\mathrm{eff}}} \tag{7.6}$$

with reduced Planck constant $\hbar$, pump mode frequency ω_0, the speed of light c, nonlinear refractive index n_2, refractive index n_0, and effective optical mode volume V_{eff}. The normalized detuning $\zeta_0 = 2\delta_\omega/\kappa$ with $\delta_\omega = \omega_0 - \omega_p$ detuning of the pump laser from the ring resonance, the normalized time $\tau = \kappa t/2$, and $d_2 = 2D_2/\kappa$ is the normalized second-order dispersion. The normalized pump amplitude is defined as:

$$f = \sqrt{\frac{8g\eta P_{\text{pump}}}{\kappa^2\hbar\omega_0}} \tag{7.7}$$

with the coupling coefficient η and the input pump power P_{pump}.

The simulation parameters used in this work are: pump wavelength $\lambda = 1553$nm, $D_1/2\pi = 200$GHz, $V_{eff} = 0.72 \times 10^{-15}$ m^3, $P_{\text{pump}} = 0.5$W to 2.25 W, $D_2 =$ 1 GHz to 2.6GHz (calculated from the simulation results shown in Fig. 7.1b), $\eta =$ 0.5 (critical coupling), $n_0 = 2.0$, $n_2 = 2.4 \times 10^{-19}$ m^2/W [18], $\kappa = 3 \times 10^8$ Hz, $n_{2\text{T}} = 2.4 \times 10^{-18}$m^2/W and $\tau_\text{T} = 10^{-8}$ s and 10^{-12} s for forward and backward tuning, respectively [13].

References

1. Geng, Y., Zhou, H., Han, X., Cui, W., Zhang, Q., Liu, B., Deng, G., Zhou, Q., Qiu, K.: Coherent optical communications using coherence-cloned Kerr soliton microcombs. Nat. Commun. **13** (2022)
2. Yang, Z., Jahanbozorgi, M., Jeong, D., Sun, S., Pfister, O., Lee, H., Yi, X.: A squeezed quantum microcomb on a chip. Nat Commun. **12** (2021)
3. Brasch, V., Lucas, E., Jost, J.D., Geiselmann, M., Kippenberg, T.J.: Self-referenced photonic chip soliton Kerr frequency comb. Light Sci. Appl. **6**, e16202–e16202 (2017)
4. Diakonov, A., Khrizman, K., Zano, E., Stern, L.: Broadband cavity-enhanced Kerr Comb spectroscopy on chip. npj Nanophoton. **1** (2024)
5. Beloy, K., Bodine, M.I., Bothwell, T., Brewer, S.M., Bromley, S.L., Chen, J.-S., Deschênes, J.-D., Diddams, S.A., Fasano, R.J., Fortier, T.M., Hassan, Y.S., Hume, D.B., Kedar, D., Kennedy, C.J., Khader, I., Koepke, A., Leibrandt, D.R., Leopardi, H., Ludlow, A.D., McGrew, W.F., Milner, W.R., Newbury, N.R., Nicolodi, D., Oelker, E., Parker, T.E., Robinson, J.M., Romisch, S., Schäffer, S.A., Sherman, J.A., Sinclair, L.C., Sonderhouse, L., Swann, W.C., Yao, J., Ye, J., Zhang, X.: Frequency ratio measurements at 18-digit accuracy using an optical clock network. Nature **591**, 564–569 (2021)
6. Kippenberg, T.J., Gaeta, A.L., Lipson, M., Gorodetsky, M.L.: Dissipative Kerr solitons in optical microresonators. Science (New York, N.Y.) **361** (2018)
7. Herr, T., Brasch, V., Jost, J.D., Wang, C.Y., Kondratiev, N.M., Gorodetsky, M.L., Kippenberg, T.J.: Temporal solitons in optical microresonators. Nature Photon **8**, 145–152 (2014)
8. Chembo, Y.K.: Kerr optical frequency combs: theory, applications and perspectives. Nanophotonics **5**, 214–230 (2016)
9. Jang, J.K., Okawachi, Y., Zhao, Y., Ji, X., Joshi, C., Lipson, M., Gaeta, A.L.: Conversion efficiency of soliton Kerr combs. Opt. Lett. **46**, 3657–3660 (2021)
10. Zhang, X., Wang, C., Cheng, Z., Hu, C., Ji, X., Su, Y.: Advances in resonator-based Kerr frequency combs with high conversion efficiencies. npj Nanophoton. **1** (2024)
11. Coen, S., Erkintalo, M.: Universal scaling laws of Kerr frequency combs. Opt. Lett. **38**, 1790–1792 (2013)

12. Gasmi, E., Peng, H., Koos, C., Reichel, W.: Bandwidth and conversion-efficiency analysis of Kerr soliton combs in dual-pumped resonators with anomalous dispersion. Phys. Rev. A **108** (2023)
13. Sabouri, S., Mendoza Velasco, L.A., Jamshidi, K.: Efficiency comparison for frequency comb formations in a silicon nitride microring within anomalous dispersion regime. Opt. Continuum **1**, 2100 (2022)
14. Guo, H., Karpov, M., Lucas, E., Kordts, A., Pfeiffer, M.H.P., Brasch, V., Lihachev, G., Lobanov, V.E., Gorodetsky, M.L., Kippenberg, T.J.: Universal dynamics and deterministic switching of dissipative Kerr solitons in optical microresonators. Nature Phys **13**, 94–102 (2017)
15. Gärtner, J., Trocha, P., Mandel, R., Koos, C., Jahnke, T., Reichel, W.: Bandwidth and conversion efficiency analysis of dissipative Kerr soliton frequency combs based on bifurcation theory. Phys. Rev. A **100** (2019)
16. Xue, X., Wang, P.-H., Xuan, Y., Qi, M., Weiner, A.M.: Microresonator Kerr frequency combs with high conversion efficiency. Laser Photon. Rev. **11** (2017)
17. Mendoza-Velasco, L.A.: Lugiato-Lefever Equation Solver. https://github.com/LuisAngelMendozaVelasco/Lugiato-Lefever_Equation_Solver. Last accessed 20 May 2024
18. Ikeda, K., Saperstein, R.E., Alic, N., Fainman, Y.: Thermal and Kerr nonlinear properties of plasma-deposited silicon nitride/silicon dioxide waveguides. Opt. Expr. **16**, 12987–12994 (2008)

Chapter 8
Coherence Scrambling Using an Integrated Optical Phased Array and a Multi-Mode Fiber

Bernat Molero Agudo, Irwan D. Setija, and Martijn J. R. Heck

Abstract Speckle noise degrades high-resolution imaging by obscuring essential details. We propose a compact, non-mechanical solution using a one-dimensional optical phased array integrated in an indium phosphide photonic integrated circuit edge-coupled to a multi-mode fiber. This system significantly reduces speckle contrast and ensures a near-uniform intensity profile by actively scrambling spatial coherence. Our experiments agree with simulated predictions and show that polarization mixing and camera resolution further decrease speckle effects. Thus, this integrated approach offers a scalable and vibration-free method for applications in metrology, holography, and other fields of coherence-based imaging. Future work will assess the performance of different optical phased array designs, as well as investigate additional ways of coherence scrambling affecting the measurements.

8.1 Introduction

Coherent light has long enabled high-resolution techniques in fields such as wafer metrology, optical coherence tomography, and holography, owing to its high brightness and well-defined phase relationships [1, 2]. Despite these advantages, coherence inherently introduces speckles—granular interference phenomena that degrades image contrast and obscures crucial details [3, 4]. Over the years, various post-processing [5], polarization mixing [6], and wavelength-diversity approaches [7] have been proposed to mitigate speckles. Although effective to some degree, these methods often add complexity, can be computationally costly, or fail to deliver the ultra-low speckle contrast demanded by cutting-edge applications [8, 9].

B. Molero Agudo (✉) · I. D. Setija · M. J. R. Heck
Department of Electrical Engineering, Eindhoven University of Technology, Eindhoven, The Netherlands
e-mail: b.molero.agudo@tue.nl

I. D. Setija
Department of Research, ASML Netherlands B.V., Veldhoven, The Netherlands

P. Smowton and T. Krauss (eds.), *Proceedings of the 26th European Conference on Integrated Optics*, Springer Proceedings in Physics 440,
https://doi.org/10.1007/978-3-032-29673-3_8

A different strategy involves scrambling the spatial coherence before imaging. Traditionally, moving diffusers or vibrating screens have been employed for this purpose [10, 11], but these mechanical elements can introduce vibrations, drift, or system bulkiness. Seeking a stable, compact alternative has led researchers to photonic integration platforms, where on-chip components allow optical beam shaping and coherence management without moving parts [12, 13]. In particular, optical phased arrays (OPAs) fabricated in indium phosphide (InP) or silicon provide agile phase control across many emitters [14–16], enabling carefully configured interference states that can significantly reduce speckles.

Here, we harness a one-dimensional (1D) OPA edge-coupled to a multi-mode fiber (MMF) as an efficient means of coherence scrambling. By actively modulating the phase in each waveguide channel, we demonstrate a robust speckle-suppression approach simpler than fully two-dimensional OPA antennas [17, 18], yet capable of delivering near-uniform illumination. Unlike many 2D OPAs that rely on vertical grating couplers and thus exhibit narrow-band behavior, our 1D edge-emitting architecture inherently supports broader bandwidths, making it more flexible for multi-wavelength metrology and imaging. In this paper, we compare our experimental results with numerical simulations and show that partial polarization and camera resolution can further accelerate speckle reduction.

8.2 Materials and Methods

We employ an on-chip OPA, a step-index MMF, and both experimental and numerical tools to quantify speckle suppression. First, we detail the photonic integrated circuit (PIC) layout and experimental setup. Then, we introduce our numerical procedure for simulating speckle formation. Finally, we define the metrics for evaluating local and global beam uniformity.

8.2.1 PIC and Experimental Setup

As shown in Fig. 8.1, our InP-based PIC comprises a one-dimensional OPA of $N_w = 16$ channels, each incorporating an electro-optic phase modulator (EOPM). An external laser (Keysight 81980A) at wavelength $\lambda = 1500\,\text{nm}$ and linewidth $\Delta\nu = 100\,\text{kHz}$ is coupled to the chip. A 1×16 splitter distributes optical power equally to each channel, allowing independent phase control.

Each channel is butt-coupled to a 1 m length step-index MMF (Thorlabs M127L01, 400 μm core, NA = 0.5), where multiple guided modes interfere to form a speckle pattern at the fiber output. As shown in Fig. 8.2, the output is collimated (Thorlabs PAF2-5C) and subsequently focused (Thorlabs LA4725-C-ML, $f = 75\,\text{mm}$)

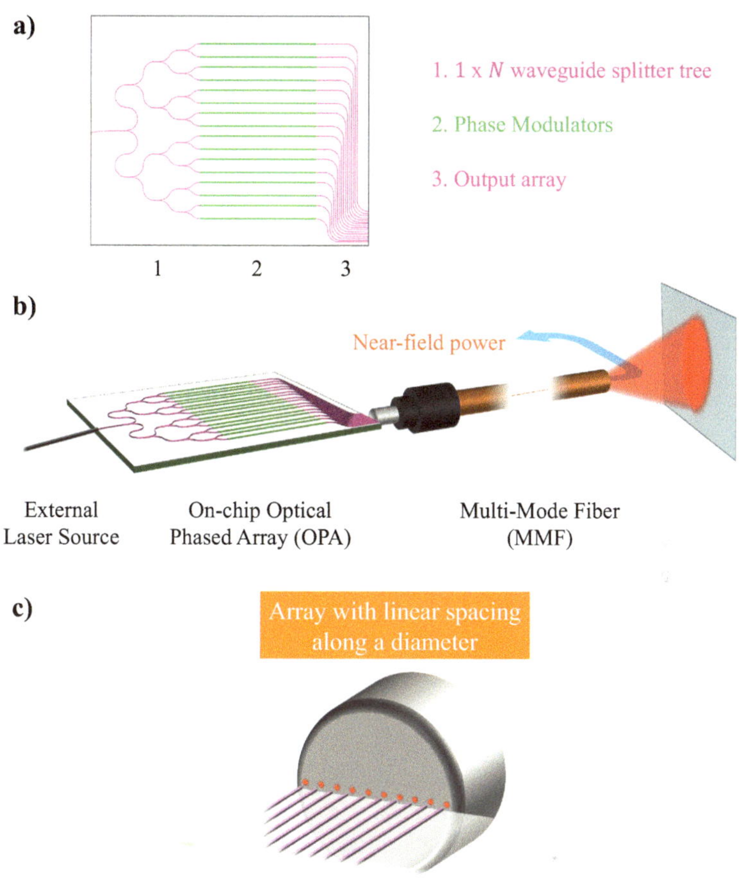

Fig. 8.1 Conceptual schematic of the on-chip optical phased array architecture. a In 1, an external laser source is coupled into a 1,×, N waveguide splitter tree on an indium phosphide photonic integrated circuit. 2 shows that each of the $N_w = 16$ output channels includes an electro-optic phase modulator for dynamic phase control. **b** The external source goes into the chip and is directly butt-coupled to a step-index multi-mode fiber. Inside this, the fields excite multiple guided modes that interfere to form speckle at the output. By randomly varying the phases on each channel, one generates independent speckle patterns that, when averaged, reduce the overall speckle contrast and yield a more uniform illumination distribution suitable for coherent imaging and metrology applications. **c** Show how the waveguides are arranged with linear radial offsets

onto an infrared camera (Xenics Xeva 320). We generate a sequence of speckle realizations by applying random phase shifts to the EOPMs. Temporal averaging of these realizations reduces the overall speckle contrast, improving illumination uniformity.

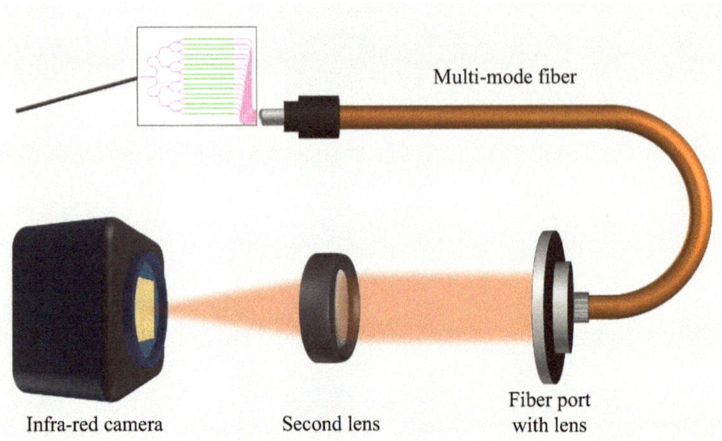

Fig. 8.2 Schematic of the experimental set-up. The on-chip 1D optical phased array (OPA) is butt-coupled to a step-index multimode fiber. The beam is collimated through a fiber port at the fiber's distal end and then focused onto an infrared camera using a plano-convex lens. Random phase settings in each OPA channel generate successive speckle patterns at the camera, enabling speckle reduction through temporal averaging

8.2.2 *Speckle Simulations*

We simulate the speckle patterns by treating each OPA channel as a distinct emitter that couples into the fiber's eigenmodes. Mathematically, the electric field at the MMF output for each speckle realization t is:

$$E_t(r,\theta) = \sum_{i=1}^{N_w} \exp(-j\,\phi_{ti}) \sum_m c_{im}\, M_m(r,\theta)\, \exp(-j\,\beta_m L),$$

where ϕ_{ti} is the random phase applied to the ith emitter in realization t, c_{im} are the coupling coefficients, $M_m(r,\theta)$ are the fiber modes, and β_m their propagation constants.

In the case of polarization diversity or finite bandwidth, a single speckle pattern is understood as the sum of N_{indep} partially incoherent patterns. We simulate this effect by rotating the fields of the waveguides before applying random phases, which is an equivalent way of getting independent patterns.

To obtain the final speckle intensity, we average over N_p random-phase realizations. This numerical framework lets us explore how speckle contrast depends on both N_p and N_{indep}, and compare these trends with our measurements.

8.2.3 Metrics for Speckle and Uniformity

We evaluate the *local* speckle variation using the contrast

$$C = \frac{\sigma(I)}{\langle I \rangle}, \tag{8.1}$$

where $\sigma(I)$ and $\langle I \rangle$ are the standard deviation and mean of the intensity over a chosen region [3]. A lower contrast implies more uniform local intensity.

To assess *global* beam uniformity, we fit the measured near-field intensity profile with Zernike polynomials up to the 7th order. The difference between the fitted profile's maximum and minimum, normalized by the mean intensity, indicates the large-scale flatness of the beam. This metric allows us to decouple macroscopic beam inhomogeneities from local speckle fluctuations.

8.3 Results and Discussion

We experimentally demonstrate our approach using a 1D OPA with *linear* spacing. Figure 8.3 summarizes the key findings, including both the uniformity of the final speckle pattern and the evolution of speckle contrast with the number of phase realizations.

Regarding global homogeneity, panel (a) of Fig. 8.3 shows the final speckle pattern obtained after summing 300 randomly phased patterns. A Zernike fit of this pattern yields a global uniformity that closely matches our simulations: 0.963 (simulation) versus 0.957 (experiment). This strong agreement validates the accuracy of our numerical model.

In terms of chip design, a linear waveguide arrangement, while theoretically suboptimal for global uniformity, can still achieve nearly flat near-field intensity. On the other hand, engineering the waveguide placement is envisioned to provide better results for homogeneity.

Regarding speckle contrast, the results shown in panel (b) of Fig. 8.3 indicate that each speckle frame already has a reduced contrast of about 0.38. Based on theory and simulation, we infer that effectively ~14 independent speckle patterns are contributing to a single frame. We attribute this partial incoherence to three factors:

- **Polarization diversity**: The input light was not polarized. Moreover, residual birefringence in the MMF may mix two orthogonal polarizations, each carrying distinct speckle structures [2]. Hence, polarization diversity adds 2 degrees of freedom out of the 14.
- **Camera resolution**: Each of the individual pixels of the camera is not small enough to capture one single speckle. Therefore, one single pixel averages over several speckle grains depending on the number of modes in the MMF and the

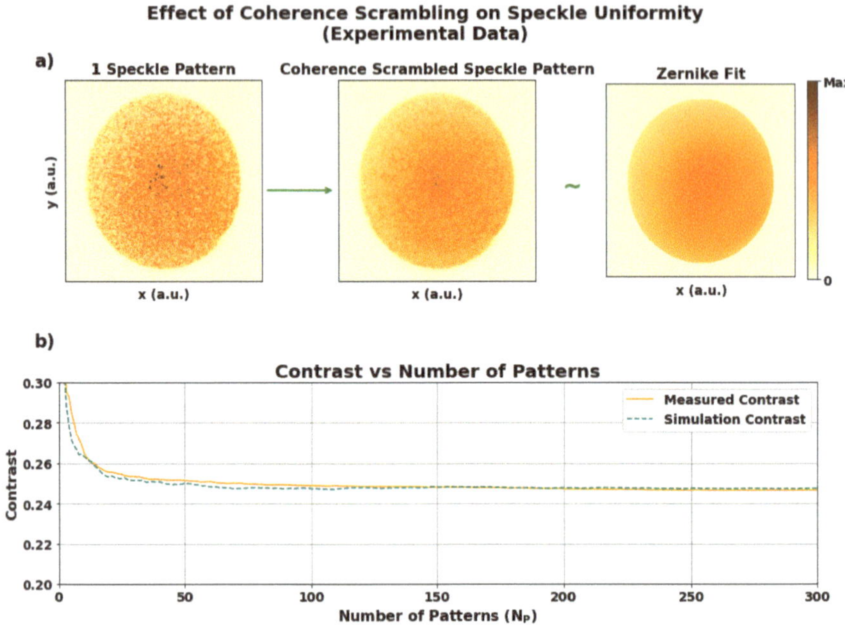

Fig. 8.3 Experimental results and comparison to simulation. a Final near-field speckle pattern after summing 300 speckle patterns by shifting the phase in the chip, and its Zernike fit (right). **b** Measured and simulated speckle contrast as a function of the number of random phase patterns summed

size of these pixels. Using the theory in [2, Sect. 8.1], the calculated degrees of freedom for contrast reduction due to resolution is 2.44.

- **Laser bandwidth**: The laser's 100 kHz linewidth leads to a finite coherence time. The finite bandwidth necessary for decorrelating speckle patterns of different wavelengths with our MMF is around 1 GHz [2], which is much larger than the laser's bandwidth. Then we conclude the bandwidth doesn't affect the contrast.

With this, we still lack approximately a factor of $14/(2.44 \times 2) \sim 3$ to fully explain the observed contrast reduction, suggesting that additional coherence-scrambling mechanisms further reduce speckles—such as mode mixing within the fiber, phase decorrelations at the OPA–fiber interface, or partial spatial coherence in the OPA output. While these sources remain to be quantified, their presence is beneficial in practice: it implies fewer waveguides or fewer temporal realizations can achieve comparable contrast suppression. Even with just 16 waveguides and 300 random phase settings, the final contrast saturates at about 0.25, which is better than purely polarized monochromatic estimates.

We note that in an industrial or clinical imaging system, there may be an even larger laser bandwidth or unintentional polarization scrambling, both of which drive the speckle contrast down. In principle, the contrast formula can be extended to account for partial coherence across the waveguides and multi-polarization states [2], although this is still part of future work.

8.4 Conclusion

We have demonstrated that coupling a one-dimensional optical phased array to a multimode fiber provides a compact, fast, broadband, and vibration-free strategy for reducing speckles. Our simulations accurately predict both contrast and large-scale beam homogeneity for $N_w = 16$ waveguides over various numbers of phase realizations N_p. Experiments confirm this model using a linear array on an indium phosphide platform and a 1 m long, NA = 0.5 MMF: the measured global uniformity (0.957) closely matches the simulated result (0.963), and the speckle contrast saturates at $C \approx 0.25$ in good agreement with simulations.

Although pushing the contrast below 1% may require more channels or additional bandwidth/polarization strategies, our proof of concept demonstrates the viability of 1D OPA plus MMF for applications in metrology, holography and coherence-based imaging. In addition, we have observed additional partial coherence losses (contrast reduction) beyond those explained by polarization mixing, wavelength diversity, and camera resolution. A critical next step is to identify and quantify these extra coherence-scrambling mechanisms so that they can be understood for even greater speckle reduction. Future work will also explore different OPA designs to improve homogeneity.

Acknowledgements This project has received support from TKI Holland High Tech under the project name PICOSTAR.

References

1. Dainty, J.C., Goodman, J.W., McKechnie, T.S., Françon, M., Ennos, A.E.: Laser Speckle and Related Phenomena, vol. 9. Springer, Berlin (1975). https://doi.org/10.1007/978-3-662-43205-1
2. Goodman, J.W.: Speckle Phenomena in Optics: Theory and Applications, 2nd edn. SPIE (2020). ISBN: 9781510631496. https://doi.org/10.1117/3.2548484
3. Goodman, J.W.: Some fundamental properties of speckle. J. Opt. Soc. Am. **66**(11), 1145–1150 (1976)
4. Rawson, E.G., Goodman, J.W.: Speckle in optical fibers. In: Applications of Speckle Phenomena, vol. 0243. SPIE, San Diego (1980). https://doi.org/10.1117/12.959282
5. Kaur, A., Dong, G.: A complete review on image denoising techniques for medical images. Neural Proces. Lett. **55**(6), 7807–7850 (2023). ISSN: 1573773X. https://doi.org/10.1007/s11063-023-11286-1
6. Ohtsubo, J.: Frequency dependence of modal noise in multimode optical fibers. Appl. Opt. **28**(19), 4235–4238 (1989)

7. Drexler, W., Fujimoto, J.G.: Optical Coherence Tomography, 2nd edn. Springer, Cham (2015)
8. Ma, R., Zhang, W.L., Guo, J.Y., Rao, Y.J.: Decoherence of fiber supercontinuum light source for speckle-free imaging. Opt. Exp. **26**(20), 26758 (2018). ISSN: 10944087. https://doi.org/10.1364/oe.26.026758
9. Reynolds, R.O., Kost, A.: A comparison of methods for the reduction of fiber modal noise in high-resolution spectrographs. In: Advances in Optical and Mechanical Technologies for Telescopes and Instrumentation, vol. 9151. SPIE (2014). ISBN: 9780819496195. https://doi.org/10.1117/12.2057024
10. Kubota, S., Goodman, J.W.: Very efficient speckle contrast reduction realized by moving diffuser device. Appl. Opt. **49**(23), 4358–4391 (2010)
11. Trisnadi, J.I.: Speckle contrast reduction in laser projection displays. Proc. SPIE **4657**, 131–137 (2002)
12. Pollock, C.R., Lipson, M.: Integrated Photonics. Springer, Boston (2003) [Online]. Available: https://doi.org/10.1007/978-1-4757-5522-0
13. Smit, M., et al.: An Introduction to InP-Based Generic Integration Technology (2014). https://doi.org/10.1088/0268-1242/29/8/083001
14. Gagino, M., Van Rijn, M.B., Bente, E.A., Smit, M.K., Dolores-Calzadilla, V.: Broadband operation of an InP optical phased array. IEEE Photon. Technol. Lett. **34**(10), 541–544 (2022). ISSN: 19410174. https://doi.org/10.1109/LPT.2022.3171979
15. Hsu, C.P., Li, B., Solano-Rivas, B. et al.: A review and perspective on optical phased array for automotive LiDAR. IEEE J. Select. Top. Quant. Electron. **27**(1) (2021). ISSN: 15584542. https://doi.org/10.1109/JSTQE.2020.3022948
16. Yi, Y., Wu, D., Kakdarvishi, V., Yu, B., Zhuang, Y., Khalilian, A.: Photonic Integrated Circuits for an Optical Phased Array (2024). https://doi.org/10.3390/photonics11030243
17. Hu, G., Zhong, K., Qin, Y., Tsang, H.K.: Silicon photonic integrated circuit for high-resolution multimode fiber imaging system. APL Photon. **8**(4) (2023). ISSN: 23780967.https://doi.org/10.1063/5.0137688
18. Bütow, J., Eismann, J.S., Sharma, V., Brandmüller, D., Banzer, P.: Generating free-space structured light with programmable integrated photonics. Nat. Photon. **18**, 243–249 (2024). ISSN: 17494893. https://doi.org/10.1038/s41566-023-01354-2

Chapter 9
Linewidth Analysis of Monolithically Integrated DBR Lasers with Different Intracavity Passive Ring Filters

Julian Konig, Rachel A. Jones, Xaveer J. M. Leijtens, and Erwin A. J. M. Bente

Abstract By integrating a passive ring resonator into the linear cavity of a DBR laser, the effective optical path length is increased, thereby extending the photon lifetime and reducing the linewidth-power product. In principle, reducing the coupling factor in and out of the ring resonator brings a significant improvement in the linewidth-power product. Measurements performed on intracavity ring lasers fabricated on a monolithic InP platform confirm this reduction in the linewidth-power product. Notably, however, while the linewidth-power product improves, the overall achievable linewidth does not. It is limited by the reduced output power associated with increased losses in the rings.

9.1 Introduction

Semiconductor lasers are essential components in coherent optical communication systems and sensing applications. Lowering the linewidth (a good measure for phase noise [1]) of these lasers can significantly decrease bit error rates and increase sensing accuracy [2, 3].

A well-known way to narrow the linewidth of a semiconductor laser is to increase the photon lifetime inside the laser cavity with a passive extension [4]. Including a passive part inside a laser cavity allows for a larger total optical field in the cavity, effectively reducing the noise contribution of spontaneous emission in the laser. The intrinsic linewidth of such a laser, $\Delta\nu$, can be calculated using a modified version of the Schawlow-Townes equation [4],

J. Konig (✉) · R. A. Jones · X. J. M. Leijtens · E. A. J. M. Bente
Photonic Integration Group, Eindhoven University of Technology, Eindhoven, The Netherlands
e-mail: j.konig@tue.nl

P. Smowton and T. Krauss (eds.), *Proceedings of the 26th European Conference on Integrated Optics*, Springer Proceedings in Physics 440,
https://doi.org/10.1007/978-3-032-29673-3_9

$$\Delta\nu = \frac{v_g^2 h\nu n_{sp}(1+\alpha^2)}{4\pi P_o}(a_m + a_i)a_m\frac{1}{C}\left(\frac{1}{1+\frac{n_{gp}L_p}{n_{ga}L_a}}\right)^2 \tag{9.1}$$

where n_{gp} and L_p are the group index and length of the passive section of the laser, and n_{ga} and L_a are the group index and length of the active section of the laser. The larger L_p is compared to L_a, the lower the linewidth. The rest of the equation is the standard Schawlow-Townes equation for a semiconductor laser with unequal mirrors, the variables are explained in detail in [5]. As can be seen from the equation, the linewidth scales linearly with output power P_o, therefore a commonly used metric is the linewidth-power product $\Delta\nu P_o$.

The method of narrowing linewidth by adding a passive section to a laser has already been implemented with great success in other work [6, 7]. As in [6], ring resonators can be used in the passive part of the extended laser cavity. This way the passive length can be long without a large footprint on chip and the ring simultaneously behaves as a wavelength filter to maintain single mode operation with a good sidemode suppression ratio [5]. Figure 9.1 shows a schematic of such an intracavity ring laser design. It contains a semiconductor optical amplifier (SOA) as a gain medium and two distributed Bragg reflectors (DBR) as mirrors and filters. The ring consists of two multimode interferometers (MMI) to couple light into and out of passive waveguide bends of 180°.

In this work, we show that by reducing the coupling coefficient of the MMIs in the rings, we can increase the effective length of the ring, and therefore we can narrow the linewidth-power product $\Delta\nu P_o$ of a laser significantly. We also report an interesting finding on the overall linewidth performance of these lasers.

9.2 Ring Transmission and Effective Length

A lower coupling factor means less light is coupled into and out of the ring. This has two conflicting effects on $\Delta\nu P_o$:

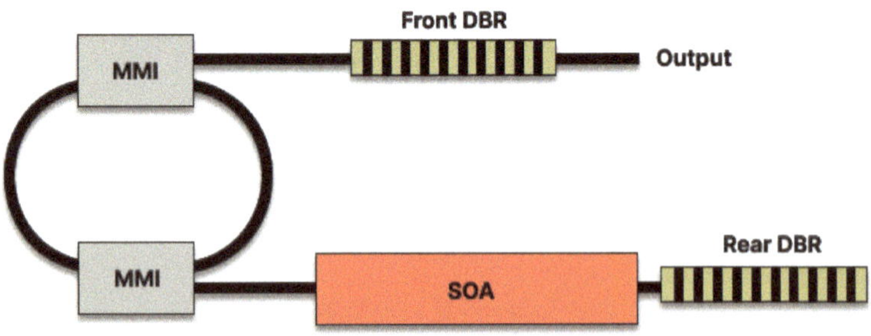

Fig. 9.1 Schematic overview of intracavity ring laser

1. The transmission through the ring is diminished, resulting in higher intracavity losses (a_i) and a higher $\Delta\nu P_\text{o}$.
2. The light undergoes more round trips resulting in a longer effective length (L_p) and a lower $\Delta\nu P_\text{o}$.

We calculate the expected transmission and effective length for the three standard MMI couplers available on the monolithic indium phosphide (InP) platform of SMART photonics [8]: 50, 28, and 15%. We assume rings of 800 μm circumference, waveguide losses of 6.5 dB/cm, and an MMI insertion loss of 0.5 dB. The results are shown in Fig. 9.2.

From the transmission and effective length, we can calculate the influence on the linewidth for the different rings using (1). The calculated power transmission of the rings, effective lengths, and resulting overall theoretical linewidth narrowing, compared to a laser without a ring, are summarized the upper half of Table 9.1.

By changing the coupling factor of the MMIs from 50 to 28%, we can achieve a further reduction in $\Delta\nu P_\text{o}$ of 0.20/0.11 = 1.82 times and by changing it from 50 to 15%, we can achieve a further reduction of around 0.20/0.07 = 2.86 times.

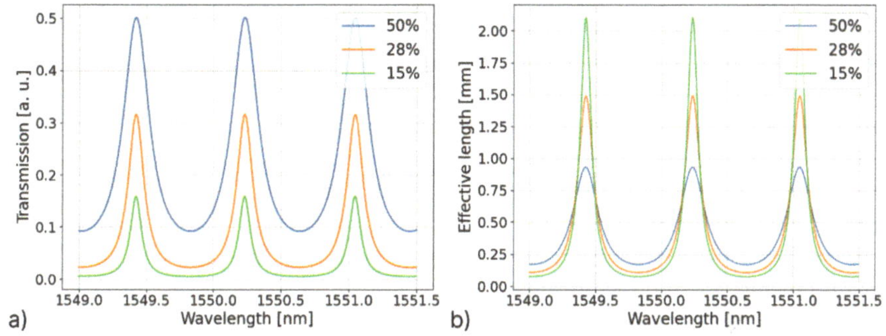

Fig. 9.2 **a** Calculated transmission of the ring resonators with different coupling factors. **b** Calculated effective length of the ring resonators with different coupling factors

Table 9.1 Calculated and measured power transmission, effective length, and resulting linewidth narrowing compared to a laser without ring for different coupling factor MMIs in rings

	Coupling factor (%)	Power transmission	Effective length (mm)	Linewidth narrowing
Calculated	50	0.50	0.93	0.20
	28	0.32	1.49	0.11
	15	0.16	2.10	0.07
Measured	50	0.424	0.80	0.24
	28	0.315	1.59	0.10
	15	0.163	2.14	0.07

9.3 Chip Design and Measurement Results

Lasers with the three different intracavity rings and test structures with only the rings were designed and fabricated on the monolithic InP platform of SMART photonics [8]. The rings are designed with MMIs with coupling factors of 50, 28, and 15%. All lasers have a front DBR length of 300 μm, a rear DBR length of 400 μm, and an SOA length of 750 μm. Figure 9.3 shows a picture of the fabricated chip marking the different components.

For measurements, light is coupled into and out of the chip with lensed fibers. Electrical signals are supplied through probe needles. First, the transmission spectrum of the rings was measured with a tuneable laser and photodiode (Agilent 8163 tuneable laser and power meter chassis). The results are shown in Fig. 9.4. By fitting the transmission of the rings, we can get values for the transmission of the ring, the effective length of the ring, and the corresponding linewidth narrowing compared to a laser without the ring. These results are shown in the lower half of Table 9.2. They are in good agreement with the predicted values.

Then the ring lasers were characterised. Figure 9.5a shows output power (Agilent 8163 tuneable laser and power meter chassis) and bias voltage (Keithley 2041 source meter) as a function of bias current (Keithley 2041 source meter) for the three lasers. As expected, the lasers with lower coupling factors have less output power due to the higher losses in the cavity. The threshold currents and corresponding threshold gains for the different lasers are summarized in Table 9.2.

Finally, the linewidth (Keopsys C band EDFA and OEwaves OE400 phase noise measurement system) was measured as a function of output power (90/10 fiber splitter and Agilent 8163 power meter) at different bias currents (ILX low noise current source). Each measurement is repeated three times. Figure 9.5b shows the linewidth measurements as a function of power. The measurements were performed for bias currents between threshold and roll-off of the output power. The power in Fig. 9.5b is

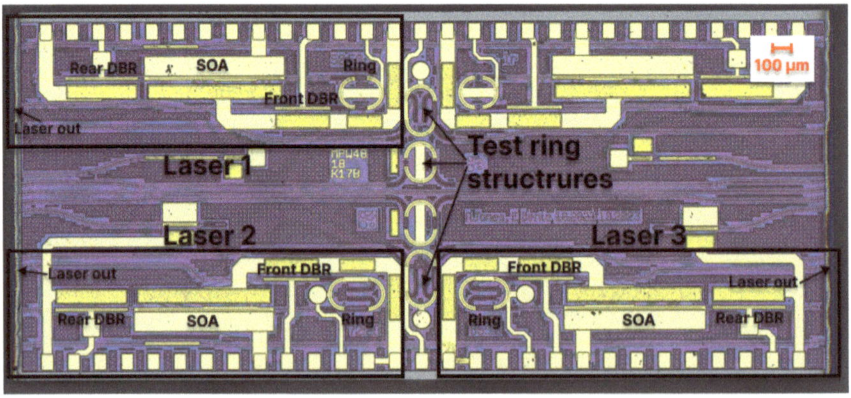

Fig. 9.3 Picture of fabricated InP chip containing test structures for the different coupling factor ring resonators and intracavity ring lasers with different coupling factors

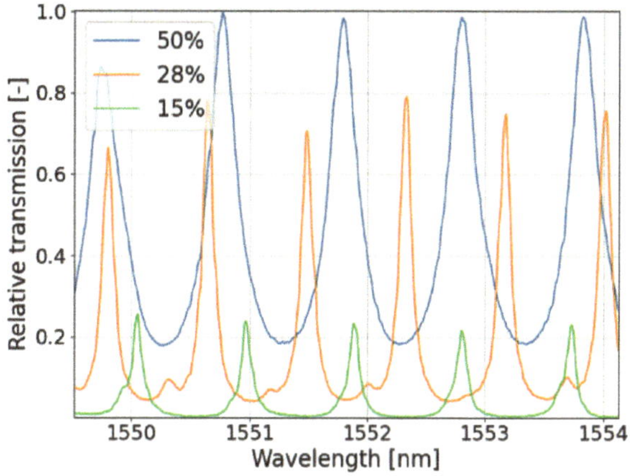

Fig. 9.4 Measured transmission spectra of the ring test structures with different coupling coefficients

Table 9.2 Threshold currents and gains for ring lasers with different coupling factor MMIs

Coupling factor (%)	Threshold current density (kA/cm^2)	Threshold gain (cm^{-1})
50	1.333	13.80
28	1.466	17.93
15	1.733	24.73

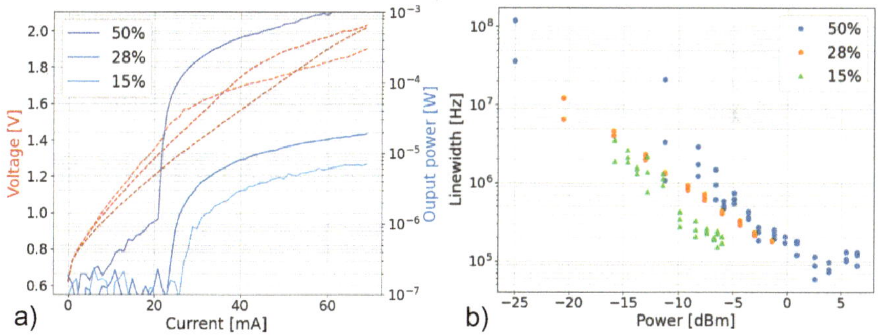

Fig. 9.5 **a** Output power and bias voltage as a function of bias current for intracavity ring lasers with different coupling factors. **b** Linewidth as a function of output power for intracavity ring lasers with different coupling factors

corrected for external losses (facet roughness, fiber losses, fiber alignment) to powers derived from the threshold gains in Table 9.2.

The change in $\Delta\nu P_o$ for the different coupling factors is clearly visible. For the 28% coupler laser, $\Delta\nu P_o$ is between 1.5 and 2 times smaller than for the 50% coupler laser, which is in good agreement with the predicted value of 0.20/0.11 = 2.2 from the measured transmission of the ring in Table 9.1. For the 15% coupler laser, $\Delta\nu P_o$ is between 3 and 5 times smaller than for the 50% coupler laser, which is also in good agreement with the 0.24/0.07 = 3.43 predicted in Table 9.1. Deviations are most likely due to variations in realized coupling factors, losses, and accuracy of the current gain relation.

9.4 Overall Achievable Linewidth

The results in Fig. 9.5b reveal an interesting finding: Although the linewidth-power product is drastically decreased, in line with predictions, the output power of the lower coupling factor ring lasers is not enough to also reach narrower overall linewidths. The minimum linewidth measured even goes up for the lower coupling factors. To reach lower overall linewidths, we need lower loss waveguides and components in the ring resonator. A good step towards this would be to develop the lasers with low loss directional couplers instead of MMIs.

9.5 Conclusion

We showed theoretically and in measurements that we can significantly decrease the linewidth-power product of intracavity ring lasers by changing the coupling factor of the rings. In our measurements we see a more than threefold decrease in $\Delta\nu P_o$ when changing the coupling coefficient of the rings from 50 to 15%.

Notably, the lower coupling factor rings induce so much losses, that even though the linewidth-power product is decreased, we do not see enough output power to reach overall lower linewidths. With the current waveguide losses and building block performance, lowering the coupling coefficient of the rings does not decrease the overall achievable linewidth in practice.

References

1. Henry, C.H.: Theory of the Linewidth of Semiconductor Lasers. IEEE J. Quantum Electron. **18**(2), 259–264 (1982)
2. Seimetz, M..: Laser linewidth limitations for optical systems with high-order modulation employing feed forward digital carrier phase estimation. In: Optical Fiber Communication Conference/National Fiber Optic Engineers Conference, OSA Technical Digest (CD). Optica Publishing Group (2008)
3. Geng, J., et al.: Narrow linewidth fiber laser for 100-km optical frequency domain reflectometry. IEEE Photonics Technol. Lett. **17**(9), 1827–1829 (2005)
4. Kazarinov, R., Henry, C.H.: The relation of line narrowing and chirp reduction resulting from the coupling of a semiconductor laser to passive resonator. IEEE J. Quantum Electron. **23**(9), 1401–1409 (1987)

5. Andreou, S., Williams, K., Bente, E.: Monolithically integrated InP-based DBR lasers with an intra-cavity ring resonator. Opt. Express **27**, 26281–26294 (2019)
6. Fan, Y., et al.: Hybrid integrated InP-Si3N4 diode laser with a 40-Hz intrinsic linewidth. Opt. Express **28**, 21713–21728 (2020)
7. Kumar, R.J., Hänsel, A., Brusatori, M.F., Nielsen, L., Augustin, L.M., Volet, N., Heck, M.J.R.: A 10-kHz intrinsic linewidth coupled extended-cavity DBR laser monolithically integrated on an InP platform. Opt. Lett. **47**, 2346–2349 (2022)
8. Augustin, L.M., et al.: InP-based generic foundry platform for photonic integrated circuits. IEEE J. Sel. Top. Quantum Electron. **24**(1), 1–10 (2018)

Chapter 10
Non-reciprocal Effects in Waveguides Based on Spin-Momentum Locking

Fernando Martín-Romero, Víctor J. Gómez, and Alberto G. Curto

Abstract Optical isolators traditionally rely on magneto-optic effects induced by externally applied magnetic fields. This study investigates the effective index contrasts induced by spin-momentum locking, which links the transverse photonic spin with the directionality of a guided mode, in non-reciprocal waveguides. These effects can be exploited in order to design all-optical isolators based on 2D semiconductors, and specifically on transition metal dichalcogenides (TMDs) integrated on a silicon nitride platform. We explore different hybrid integrated configurations, identifying the geometrical and physical conditions for a non-reciprocal behavior. Furthermore, we simulate the absorption contrast and non-reciprocal phase shift for a TE_0 mode in a waveguide covered by a TMD monolayer, investigating their dependence on the material parameters. The results discussed in this work allow the design of all-optical devices for integrated isolation, based either on a direct contrast in the transmission at both propagation directions, or through the implementation of a Mach–Zehnder interferometer that takes advantage of the non-reciprocal phase shift effect. Our analysis contributes to the development of integrated optical isolators based on spin-momentum locking without external magnetic fields.

10.1 Introduction

Optical isolators are a key component in the design of a photonic integrated circuit. They help to avoid unwanted reflections that can negatively affect the performance of devices like lasers, or generate interferences in transmission lines [1]. The working principle of an isolator relies on breaking the temporal symmetry of the circuit, generating a non-reciprocity between forward and backward propagation. The most

F. Martín-Romero (✉) · V. J. Gómez
Nanophotonics Technology Center, Universitat Politècnica de València, Valencia, Spain
e-mail: fmarrom@ntc.upv.es

A. G. Curto
Photonics Research Group, Department of Information Technology, Ghent University-Imec, Ghent, Belgium

P. Smowton and T. Krauss (eds.), *Proceedings of the 26th European Conference on Integrated Optics*, Springer Proceedings in Physics 440,
https://doi.org/10.1007/978-3-032-29673-3_10

common approach relies on magneto-optic effects like Faraday rotation, which produces a change in the plane of polarization of linearly polarized light. However, they traditionally present the significant drawback of requiring an externally applied magnetic field [2, 3].

In this work, we have explored non-reciprocal effects based on 2D semiconductors via spin-momentum locking [4, 5]. Specifically, we propose hybrid integration of TMD (transition metal dichalcogenide) materials with silicon nitride in order to achieve optically controllable magneto-optic effects. Our results are of interest for the design and implementation of all-optical integrated isolators without the requirement of an externally applied magnetic field.

10.2 Spin-Momentum Locking and Off-Diagonal Permittivity

Guided modes confined in a waveguide carry a longitudinal field component along the propagation direction, which presents a $\pi/2$ phase shift with respect to the transverse field of the mode. This circular polarization parallel to the waveguide translates to a photonic spin transverse to the propagation direction. On a given side of the waveguide, the transverse spin changes its sign when the propagation direction is reversed. Such one-to-one relationship between the transverse spin and the directionality of a guided mode is defined as spin-momentum locking [6–9].

Magneto-optic (MO) materials are known to induce polarization-dependent effects, due to the shape of their permittivity tensor. Specifically, the permittivity of a MO two-dimensional material with losses is

$$\bar{\bar{\varepsilon}} = \begin{pmatrix} \varepsilon_{xx} & \varepsilon_{xy} & 0 \\ -\varepsilon_{xy} & \varepsilon_{xx} & 0 \\ 0 & 0 & \varepsilon_{zz} \end{pmatrix}, \tag{10.1}$$

where the elements of the matrix are complex valued, and ε_{zz} is the out-of-plane component parallel to the magnetization [10, 11]. The eigenstates of the tensor correspond to right and left circularly polarized light with a permittivity of $\varepsilon_{\pm} = \varepsilon_{xx} \pm i\varepsilon_{xy}$. In the case of a TMD monolayer oriented on the xy plane, the off-diagonal permittivity can be induced by applying a circularly polarized pump beam with a photonic spin parallel to the z direction [8, 12]. The band diagram of this family of materials shows two different energy gaps, known as the K and K' valleys [13, 14]. Excitons with opposite spin can be formed on each of these valleys via pumping with left or right circularly polarized light, generating a spin imbalance and, subsequently, an all-optical magneto-optic effect near the exciton wavelength [12]. The off-diagonal permittivity induced through this mechanism depends on the pump power.

The eigenstates of a material with the permittivity tensor described in Eq. 10.1 lead to a different real and imaginary part of the refractive index for left and right circularly polarized light. In free space, it results in a handedness-dependent phase velocity and absorption coefficient, which are known respectively as the Faraday effect and circular dichroism (CD). In the case of an integrated waveguide, it translates into non-reciprocal effects due to spin-momentum locking, since forward- and backward-propagating light will present circular polarization with opposite handedness [5]. In this study, we have designed and simulated TMD covered waveguides exploiting this non-reciprocity.

10.3 Waveguide Transmission Contrast and Non-reciprocal Phase Shift

Figure 10.1a shows the design of a TMD-on-SiN non-reciprocal waveguide with a width of 450 nm and a thickness of 200 nm, supporting a fundamental guided mode at a wavelength of 612 nm (near the neutral exciton resonance of WS_2). Since the transverse spin (S_z) points in the opposite direction at the left and right side of the waveguide (Fig. 10.1b-c), the TMD monolayer must cover only one of the sides. Then, the average S_z at the position of the monolayer will change sign when the propagation direction is reversed.

The permittivity contrast at the TMD between both directions leads to a contrast on the effective index of the guided modes. On the one hand, the contrast on the imaginary part creates a different absorption for the forward- and backward-propagating light. On the other hand, the contrast on the real part leads to a different phase shift for each propagation direction (non-reciprocal phase shift).

Using COMSOL Multiphysics, we have simulated the absorption contrast in three different situations. The first one corresponds to a non-reciprocal material with $\varepsilon_{yx} = -\varepsilon_{xy}$ (e.g., an optically pumped TMD) covering the left side of a SiN waveguide. In such case, the contrast ΔT depends linearly on the real part of the non-diagonal permittivity component, $\mathrm{Re}(\varepsilon_{xy})$, as shown on Fig. 10.2. The second case consists of the same material covering the full width of the waveguide core. On average, the transverse spin at the position of the magneto-optic material will be zero. Therefore, this configuration does not generate any transmission contrast. The last case corresponds to a material with $\varepsilon_{yx} = \varepsilon_{xy}$ covering the left half of the waveguide. Such a material does not present a non-reciprocal behavior due to the symmetry of the matrix. Only the first configuration can be used to build an optical isolator based on absorption.

In an optical isolator, the insertion losses can be understood as the attenuation of light along the direction of transmission. As expected by looking at the eigenvalues of Eq. 10.1, these losses depend on both the imaginary part of the diagonal component, $\mathrm{Im}(\varepsilon_{xx})$, and the real part of the off-diagonal component, $\mathrm{Re}(\varepsilon_{xy})$ (Fig. 10.3a).

We have also simulated the non-reciprocal phase shift per unit length, which depends linearly on the imaginary part of the non-diagonal permittivity component, $\mathrm{Im}(\varepsilon_{xy})$, as well as the propagation length required to generate a π phase shift between

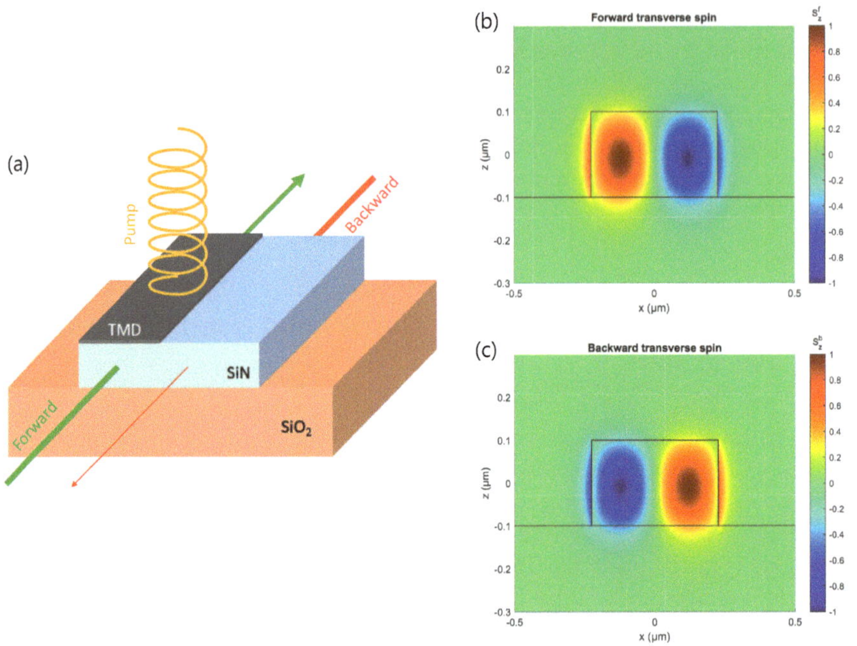

Fig. 10.1 **a** Schematic design of an optically pumped SiN waveguide covered by a TMD monolayer on its left side. **b** Forward transverse spin profile for the TE_0 mode. **c** Backward transverse spin profile for the TE_0 mode

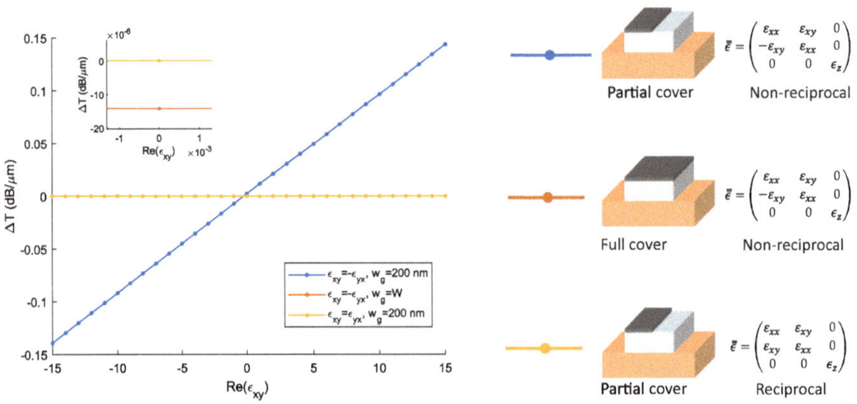

Fig. 10.2 Transmission contrast in a SiN waveguide partially covered by an optically pumped monolayer of a non-reciprocal material, fully covered by a non-reciprocal material, or partially covered by a reciprocal material. Only the first configuration results in a contrast different than zero

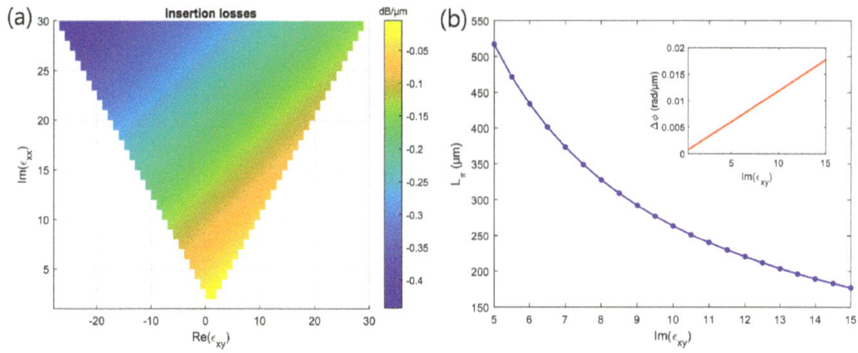

Fig. 10.3 **a** Insertion losses in a SiN waveguide covered by an optically pumped TMD monolayer acting as an optical isolator, as a function of Im(ε_{xx}) and Re(ε_{xy}). **b** Non-reciprocal phase shift per unit length (inset) and propagation length necessary to induce a π phase shift, in the same TMD covered waveguide

both directions (Fig. 10.3b). This effect can be used to build an optical isolator based on a Mach–Zehnder interferometer [15].

10.4 Conclusion

In this study, we have investigated the potential of spin-momentum locking in waveguides covered by 2D semiconductors, specifically transition metal dichalcogenides (TMDs), for the design of all-optical integrated isolators. By optically pumping the TMD, a non-reciprocal behavior can be induced in the waveguide. There appears a contrast between the effective index of the guided modes propagating along the forward and backward direction, which leads to a transmission contrast and a non-reciprocal phase shift. Our simulations show the dependence of these non-reciprocal effects on the off-diagonal permittivity of the 2D semiconductor, as well as the required geometry of the hybrid waveguide.

Ultimately, this work contributes to the growing field of non-reciprocal photonic devices based on 2D materials, providing new insights into the integration of TMDs with silicon nitride waveguides to enable all-optical isolation in the absence of external magnetic fields. Further optimization of material properties and device structures could lead to the realization of high-performance, scalable optical isolators suitable for a wide range of integrated photonic systems.

References

1. Jalas, D., Petrov, A., Eich, M., Freude, W., Fan, S., Yu, Z., Baets, R., Popović, M., Melloni, A., Joannopoulos, J.D., Vanwolleghem, M., Doerr, C.R., Renner, H.: What is—and what is not—an optical isolator. Nature Photon. **7**, 579–582 (2013)
2. Bi, L., Hu, J., Jiang, P., Kim, D.H., Dionne, G.F., Kimerling, L.C., Ross, C.A.: On-chip optical isolation in monolithically integrated non-reciprocal optical resonators. Nature Photon. **5**, 758–762 (2011)
3. Ren, S.-Y., Yan, W., Feng, L.-T., Chen, Y., Wu, Y.-K., Qi, X.-Z., Liu, X.-J., Cheng, Y.-J., Xu, B.-Y., Deng, L.-J., Guo, G.-C., Bi, L., Ren, X.-F.: Single-photon nonreciprocity with an integrated magneto-optical isolator. Laser Photonics Rev. **16**, 2100595 (2022)
4. Lodahl, P., Mahmoodian, S., Stobbe, S., Rauschenbeutel, A., Schneeweiss, P., Volz, J., Pichler, H., Zoller, P.: Chiral quantum optics. Nature **541**, 473–480 (2017)
5. Ma, J., Xi, X., Yu, Z., Sun, X.: Hybrid graphene/silicon integrated optical isolators with photonic spin–orbit interaction. Appl. Phys. Lett. **108**, 151103 (2016)
6. Ma, Y., Zhao, H., Liu, N., Gao, Z., Mohajerani, S.S., Xiao, L., Hone, J., Feng, L., Strauf, S.: On-chip spin-orbit locking of quantum emitters in 2D materials for chiral emission. Optica. **9**, 953–958 (2022)
7. Bliokh, K.Y., Bekshaev, A.Y., Nori, F.: Extraordinary momentum and spin in evanescent waves. Nat. Commun. **5**, 3300 (2014)
8. Gong, S.-H., Alpeggiani, F., Sciacca, B., Garnett, E.C., Kuipers, L.: Nanoscale chiral valley-photon interface through optical spin-orbit coupling. Science **359**, 443–447 (2018)
9. Shreiner, R., Hao, K., Butcher, A., High, A.A.: Electrically controllable chirality in a nanophotonic interface with a two-dimensional semiconductor. Nat. Photon. **16**, 330–336 (2022)
10. Zhao, D., Höchst, H., Huber, D.L.: Dielectric tensor formulation of magnetic dichroism sum rules. J. Appl. Phys. **84**, 2858–2860 (1998)
11. Sato, K., Ishibashi, T.: Fundamentals of magneto-optical spectroscopy. Front. Phys. **10**, 946515 (2022)
12. Hao, K., Shreiner, R., Kindseth, A., High, A.A.: Optically controllable magnetism in atomically thin semiconductors. Sci. Adv. **8**, eabq7650 (2022)
13. Godiksen, R.H., Wang, S., Raziman, T.V., Gómez Rivas, J., Curto, A.G.: Impact of indirect transitions on valley polarization in WS_2 and WSe_2. Nanoscale. **14**, 17761–17769 (2022)
14. Plechinger, G., Nagler, P., Arora, A., Schmidt, R., Chernikov, A., del Águila, A.G., Christianen, P.C.M., Bratschitsch, R., Schüller, C., Korn, T.: Trion fine structure and coupled spin–valley dynamics in monolayer tungsten disulfide. Nat. Commun. **7**, 12715 (2016)
15. Yokoi, H.: Calculation of nonreciprocal phase shift in magneto-optic waveguides with Ce:YIG layer. Opt. Mater. **31**, 189–192 (2008)

Chapter 11
Development of a Double Flip Fusion Bonding Approach for GaN-On-Si Heterogenous Integration

Tarni Aggarwal, Marcus Dornetshumer, Maria Mayr, Martijn Heck, and Yuqing Jiao

Abstract We have developed a double flip bonding approach to transfer a GaN laser epitaxial membrane on a pre-processed Si wafer, keeping the original orientation (n-to-p) of the epitaxial layer as on the growth substrate. We introduced an adhesive polymer based temporary bonding which can withstand the backside processing of GaN substrate removal. We have engineered each step individually, and also demonstrate fusion bond between Fs-GaN substrate and Si. The findings in this work greatly influence the development of GaN-on-Si heterogenous platform for future visible PICs.

11.1 Introduction

Photonic integration circuits (PICs) enable on-chip scaling of photonic components. PICs operating in visible spectrum (300–700 nm) are gaining interest due to exciting applications in biosensing [1–3], quantum computing [4–6], AR/VR display [7, 8]. A range of dielectric materials exhibit low propagation losses in the visible, this includes silicon nitride (Si_3N_4), aluminum oxide (Al_2O_3) [9], aluminum nitride (AlN) [10], and titanium oxide (TiO_2) [11]. In order to integrate these dielectric waveguides (WG) with a gain medium that emits visible light, hybrid or heterogeneous integration is a viable approach [12]. Most of the visible lasers utilize GaN as the gain medium. A typical laser stack consists of InGaN/GaN based multiple quantum wells with waveguiding and cladding layers grown on sapphire or a free-standing (FS) GaN substrates [13, 14].

T. Aggarwal (✉) · M. Heck · Y. Jiao
Eindhoven University of Technology, Eindhoven, The Netherlands
e-mail: t.aggarwal@tue.nl

M. Dornetshumer · M. Mayr
EV Group, Sankt Florian ann Inn, Austria

P. Smowton and T. Krauss (eds.), *Proceedings of the 26th European Conference on Integrated Optics*, Springer Proceedings in Physics 440,
https://doi.org/10.1007/978-3-032-29673-3_11

Although, hybrid integration of AlGaInN lasers butt-coupled to SiN, are demonstrated in near-UV, with an insertion loss of ~ 7.5 dB [15], it requires precise alignment and incur high packaging cost. In contrast, on-chip heterogeneous integration offers improved alignment accuracy and a cost-effective solution for mass production. While progress has been made in heterogeneous near-infrared III–V lasers [16–19], realizing heterogenous laser in visible remains challenging due to technological limitations in processing and epitaxial growth of GaN [20]. Typically used techniques such as micro-transfer printing (MTP) [21] and flip-chip die-to-wafer (D2W) bonding [22] utilize a pre-processed laser dies to integrate it with the passive platform. However, narrow WG dimensions (< 300 nm) in the visible spectrum makes it difficult to achieve precise alignment of the laser dies with the passive WG [3]. In contrast, unprocessed epitaxial wafer bonding can offers precise nanometer-range alignment accuracy through lithography [18, 19]. However, direct wafer-to-wafer (W2W) transfer of the epitaxial stack onto a pre-processed Si wafer results in an inverted laser stack relative to its orientation on the growth substrate (i.e. n-to-p). This configuration leads to an inefficient light coupling between GaN-core and the passive WGs, due to intermediate p-AlGaN layer which incur high propagation losses, stem from free carrier absorption. One way to circumvent this is to grow a reverse epitaxial stack from p-to-n. However, p-dopant activation inside a buried p-layer in GaN is difficult due to low diffusivity of the byproducts such as H^+ ions through top the n-AlGaN layer [23].

In order to transfer the laser epitaxial layer in its original orientation (i.e. n-to-p), we introduce a double flip adhesive bonding approach, which involves a temporary transfer of the laser stack onto a carrier wafer followed by the GaN substrate removal and a subsequent permanent bonding to the pre-processed Si wafer. Au-In based solid liquid interdiffusion (SLID) temporary bonding of GaN-on-glass was demonstrated [20], followed by a photoelectrochemical (PEC) undercut etch in a KOH solution. Such bonding, however, requires high de-bonding temperature (> 200 °C) and need small GaN islands for PEC etching [20].

In this work, we demonstrate a polymer based temporary bonding, owing to its advantages such as low bonding temperature (< 100 °C), stress-free bonded membrane, lower total thickness variation (TTV), low cost, and compatibility with standard cleanroom processes [24]. We aim to develop a temporary bond which can withstand the backside processing while enabling easy debonding. Instead of PEC etching, we developed a grinding, polishing, and chemical mechanical polishing (CMP) based process to remove the FS-GaN substrate from a full 50-mm wafer and prepare a bonding ready surface. We have individually engineered each step of the double flip bonding, including oxide-oxide fusion bond in the final step. We have successfully bonded a FS-GaN wafer on a Si wafer using inductively coupled plasma enhanced chemical vapor deposition technique (ICP-PECVD) to achieve smooth SiO_2 surface for fusion bonding. The developments presented in this work is crucial for the advancement of a GaN-on-Si heterogenous platform.

11.2 Double Flip Bonding of GaN on Si

A step-by-step process of the double flip bonding is shown in Fig. 11.1. We aim transfer a laser epitaxial membrane grown on a FS-GaN substrate (wafer 1) to a pre-processed Si wafer (wafer 2) with passive WGs and SiO_2 cladding, planarized by CMP. In the first step, wafer 1 is temporarily bonded to a carrier wafer (details in Sect. 2.1). Step 2 involves thinning and removing the FS-GaN substrate with CMP process (see Sect. 2.2 for details). In Step 3, the exposed n-AlGaN membrane is cleaned and covered with a low-temperature ICP-PECVD SiO_2, achieving surface roughness suitable for fusion bonding (see Sect. 2.3 for details). In Step 5, the carrier wafer is released, followed by surface cleaning. Each step of this process is optimized individually and discussed in the following section.

11.2.1 Temporary Bonding with Carrier Wafer

To develop a temporary bonding for the double flip, the selection of a suitable adhesive polymer and the associated de-bonding method is crucial. The polymer must enable bonding and de-bonding at suitable temperatures to prevent damage to the epitaxial stack and it should be mechanically stable during backside grinding and polishing. Commonly used epoxy or thermoplastic polymers rely on thermal and mechanical slide-off for de-bonding. However, this approach risks in exceeding the mechanical stress and temperature limits of de-bonding during backside grinding, which may lead to unintentional slide-off. Thick thermoplastic polymer also results in high TTV

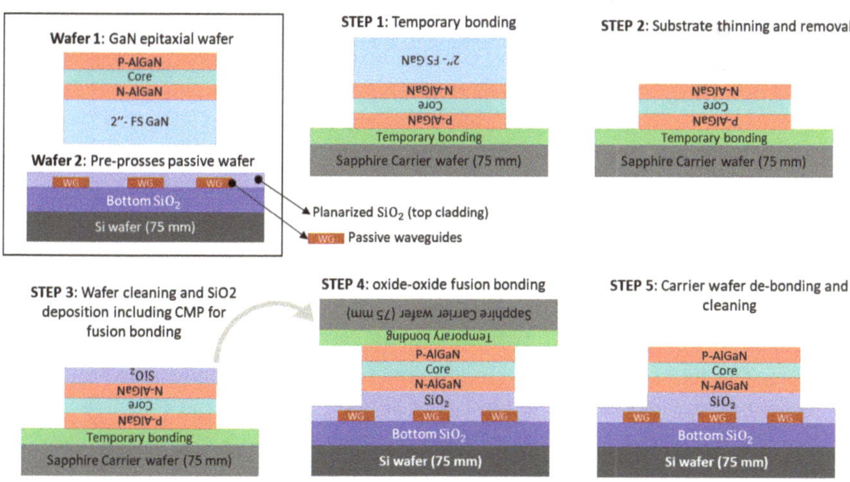

Fig. 11.1 Schematic of a double flip bonding approach for a GaN-on-Si heterogenous platform with low loss dielectric waveguides

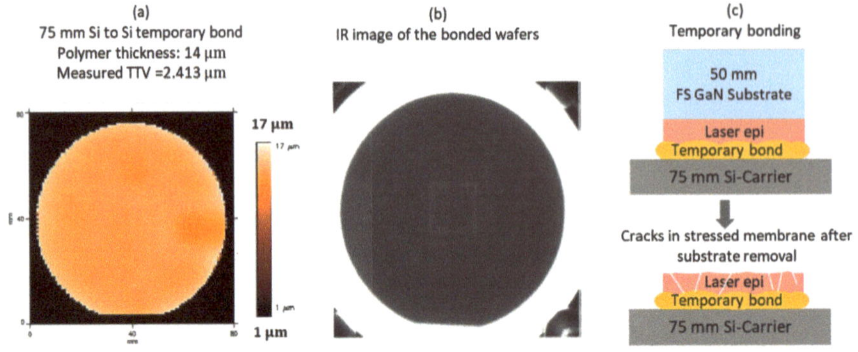

Fig. 11.2 **a** The wafer map, **b** IR image of the Si–Si temporary bond with a thick, the white square in the center of IR image is a reflection of the camera. **c** The schematic of the membrane cracking due to high TTV

across the wafer. In order to evaluate the TTV of such thermoplastic bond, we tested a Si to Si bond using a ~ 14 µm thick polymer, because Si wafer are flat (5–8 µm bow for 75-mm wafers) and cheap compared to FS-GaN substrates. Figure 11.2 shows the 2(a) wafer map and 2(b) IR image of the Si–Si temporary bond. The measured TTV is ~ 2.431 µm with 3 mm edge exclusion. Additionally, the incoming bow of the FS-GaN wafer (10–15 µm for 50-mm wafer) would likely have a negative influence on the TTV. Higher TTV and mechanical stiffness of the temporary bond can lead cracks and blistering in stretched GaN-epitaxial layers during grinding], as depicted in Fig. 2c [25, 26].

In order to achieve better TTV and robust de-bonding mechanism, we developed a laser de-bonding approach at EV groups in Austria, using a ultraviolet light (UV) cured laser release layer (LRL). This will eliminate the risk of unintentional slide-off, as de-bonding can only triggered by UV light exposure in this case. The schematic of the adhesive bonding process is shown in Fig. 11.3 for a FS-GaN bonded onto a transparent carrier. The first step is wafer cleaning in deionized (DI) water with spin dry. The second step is to spin a LRL of ~ 250 nm on the carrier wafer and a BCB polymer layer of ~ 3.5 µm on the GaN wafer using EVG®101 advanced spin coating system, followed by the soft bake on EVG®105 table-top hotplate. The bonding is established under high vacuum conditions (1×10^{-5} mbar) using EVG®520-IS bonding system. The bonding force is optimized such that it is strong enough to erase the most significant thickness variations but weak enough to prevent the outflow of the polymer through the edges [27]. Finally, the bonded wafer pair is cured under UV light. The first bond test is performed on a 75-mm Si wafer bonded on quartz carrier. The wafer map and a camera image of the bonded pair is shown in Fig. 4a and b, respectively. The measured TTV is 0.631 µm which is a significantly less than the TTV achieved with ~ 14 µm thick thermoplastic polymer in Fig. 2a. We applied the same adhesive bonding recipe (in Fig. 11.3) on a FS-GaN substrate. In Fig. 4c the wafer map of a 50-mm FS GaN substrate bonded onto a 75-mm quartz carrier is depicted. The TTV is measured TTV is 1.314 µm is achieved. This value is higher

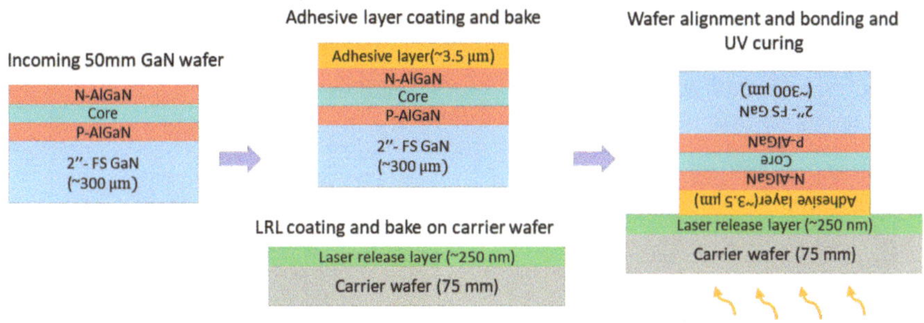

Fig. 11.3 Schematic of the proposed adhesive bonding process of FS-GaN on a quartz carrier, utilizing a laser release layer for de-bonding

than the Si-Quartz bonding due to incoming wafer bow of the FS-GaN substrate. However, further optimization of the bonding parameters such as, polymer thickness, bonding force, annealing temperature, and the use of a similar CTE (coefficient of thermal expansion) matched carrier like sapphire, could enhance bonding quality and TTV [24, 25]. After several optimizations, we have successfully demonstrated a defect-free bond between a 50-mm laser epitaxial grown on FS-GaN substrate and a 100 mm sapphire carrier wafer. The camera image of the bond is shown in Fig. 4d. The laser stack is ~ 2 μm thick and corresponds to a 450 nm of emission wavelength.

11.2.2 Substrate Thinning by Grinding and Polishing

FS-GaN substrates are optimal for defect-free growth of laser epitaxial layers [28], nevertheless, it is difficult to remove the GaN substrate by standard wet and dry etch process due its physical properties [29]. Therefore, we utilize the combination of grinding, polishing, and CMP for substrate thinning and removal to achieve bonding-ready smooth surface [30]. To test if the adhesive bonding in Fig. 4c can endure mechanical forces during grinding, the GaN-Quartz bonded pair underwent coarse and fine grinding process with diamond abrasives of different sizes ranging from 10^2 to 10^{-1} μm. The obtained material removal rate (MRR) during coarse and fine grinding is about 5–6 μm/s and 0.2–1.0 μm/s, respectively. Which helps us to obtain the thickness tolerance of 10 to12 μm during grinding. The root mean square (RMS) roughness after fine grinding was ~ 10 nm which can be brought down to < 1 nm after CMP of about 5 μm of GaN layer. The camera image of the thinned and polished GaN substrate is depicted in Fig. 11.5 along with optical microscope image showing intact wafer edges after extensive grinding and CMP. We able to remove ~ 154 μm of the GaN out of 300 μm FS-GaN substrate, which can be further reduced to 5–10 μm by grinding and further 5 μm by CMP to remove the entire substrate and obtain the bonding ready substrate. Nevertheless, the propose adhesive bonding with LRL is proved to be suitable for GaN substrate removal.

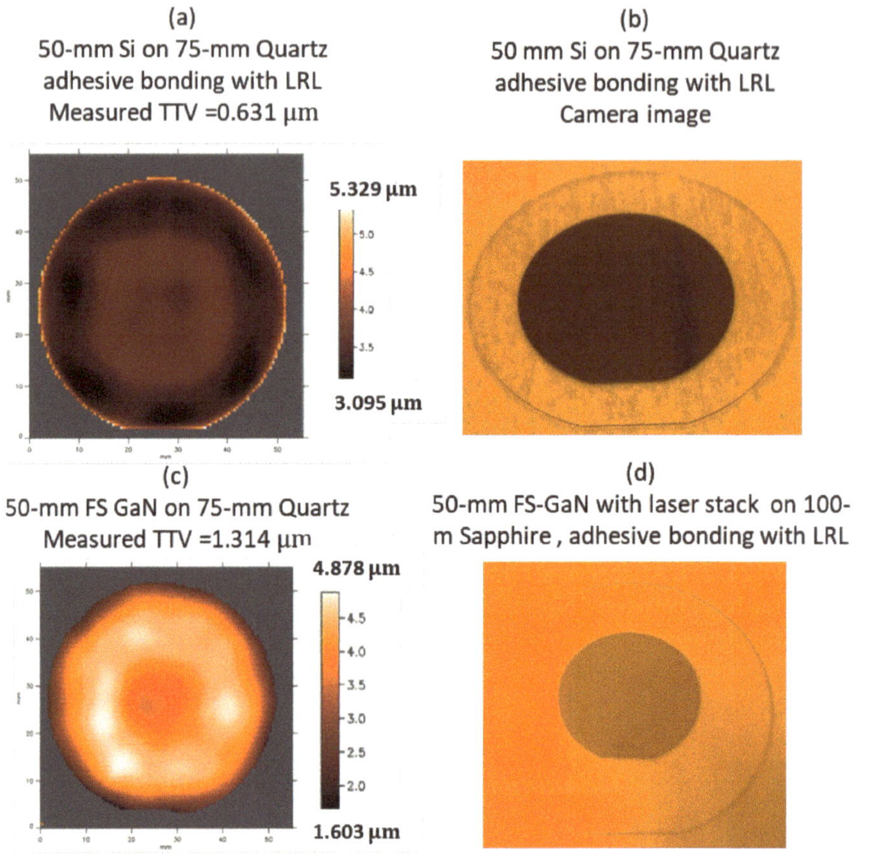

Fig. 11.4 **a** Wafer map, **b** camera image of the adhesive bonding test on Si and quartz wafers. The same bond process performed on **c** FS-GaN and quartz wafer **d** FS-GaN with laser epitaxial bonded on sapphire carrier

11.2.3 Oxide-Oxide Fusion Bonding of FS-GaN on Si Substrate

After substrate removal, the final step is to prepare the surfaces of the GaN wafer and the Si wafer for fusion bonding (step 3 and 4 in Fig. 11.1). The main challenge is to ensure the low surface roughness (< 0.5 nm) of the oxide layer on both wafers. In order to achieve smooth SiO_2, we employed low-temperature ICP-PECVD deposition technique. Figure 6a shows the AFM scan of 100 nm SiO_2 deposited on FS-GaN at 80 °C, and we have achieved the root mean square (RMS) roughness of ~ 0.54 nm. We tested fusion bond of FS-GaN substrate (with 100 nm ICP-PECVD oxide) onto a Si wafer with 200 nm thermal oxide (having surface roughness of < 0.5 nm). The bonding process starts with wafer cleaning and surface plasma activation by EVG®301 semi-automated single wafer cleaning system and EVG®810 LT

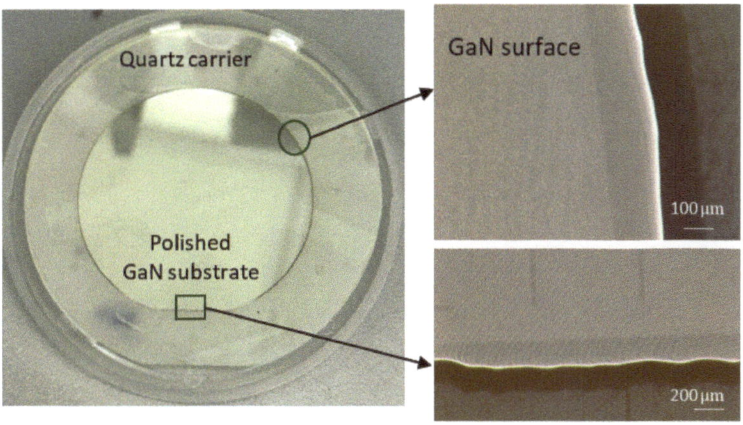

Fig. 11.5 **a** Camera image of the GaN membrane stick to the quartz wafer after grinding, polishing, and CMP of the bond pair in Fig. 4c. The intact edges and polished GaN surface are also depicted in the images taken in optical microscope

Low Temp® Plasma Activation System, respectively. The bonding is conducted in EVG®520-IS system under vacuum with a force of over 600 N, followed by annealing at 150–200 °C. Figure 6b depicts the camera image of the bonded pair. The dark part is the bonded area and the white spots are the unbonded patches. The unbonded area at the edges is essentially caused by the incoming wafer bow of FS-GaN substrate. The voids are mainly due to particles remained on the wafer surface. However, the bonding can be improved further by optimizing force and annealing during bonding. Nevertheless, the ICP-PECVD oxide is the good alternative to eliminate the CMP step for roughness improvement in our double flip bonding.

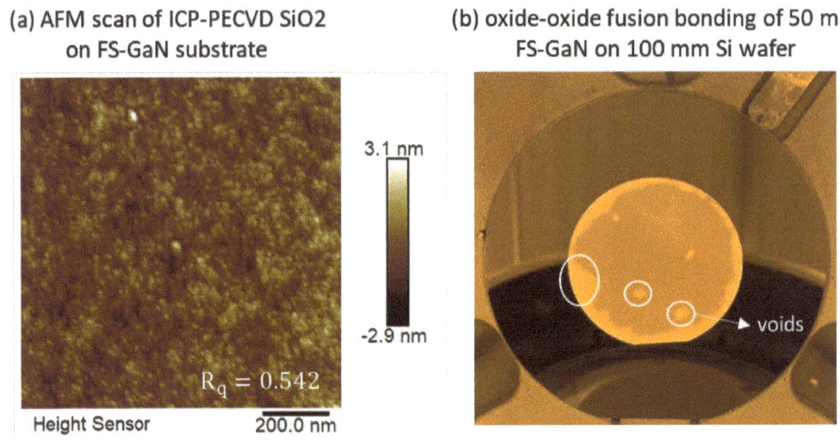

Fig. 11.6 **a** AFM scan of ICP-PECVD deposited 100 nm silicon oxide on FS-GaN substrate, rems roughness of 0.542 nm has been achieved. **b** Camera image of the fusion bond pair of FS-GaN substrate and 100 mm Si wafer with 200 nm thermal oxide

11.3 Conclusion

In conclusion, we proposed a double flip bonding method for integrating GaN-based lasers on Si substrates. Key steps include optimizing a temporary bonding process that endures extensive backside grinding and polishing. Our adhesive bonding method, featuring a laser release layer, ensures improved thickness variation and enhances mechanical and thermal stability during grinding. Additionally, we successfully achieved fusion bonding using ICP-PECVD, eliminating the need for chemical–mechanical polishing (CMP). This helps us in identifying the surface roughness and TTV requirements for the final fusion bond after substrate removal. These efforts significantly contribute towards the development of a GaN-based heterogeneous platform on Si, leveraging the low-loss dielectric waveguides.

References

1. Mohanty, A., et al.: Reconfigurable nanophotonic silicon probes for sub-millisecond deep-brain optical stimulation. Nat. Biomed. Eng. **4**, 223–231 (2020)
2. Biophotonic sensors with integrated Si_3N_4-organic hybrid (SiNOH) lasers for point-of-care diagnostics. Light Sci. Appl. https://www.nature.com/articles/s41377-021-00486-w
3. Sacher, W.D., et al.: Visible-light silicon nitride waveguide devices and implantable neurophotonic probes on thinned 200 mm silicon wafers. Opt. Express **27**, 37400–37418 (2019)
4. 2022 Roadmap on integrated quantum photonics—IOP Science. https://doi.org/10.1088/2515-7647/ac1ef4
5. Integrated optical multi-ion quantum logic. Nature. https://www.nature.com/articles/s41586-020-2823-6

6. Niffenegger, R.J., et al.: Integrated multi-wavelength control of an ion qubit. Nature **586**, 538–542 (2020)
7. Shin, M.C., et al.: Chip-scale blue light phased array. Opt. Lett. **45**, 1934–1937 (2020)
8. Raval, M., Yaacobi, A., Watts, M.R.: Integrated visible light phased array system for autostereoscopic image projection. Opt. Lett. **43**, 3678–3681 (2018)
9. Versatile silicon nitride and alumina integrated photonic platforms for the ultraviolet to short-wave infrared. IEEE J. Mag. IEEE Xplore. https://ieeexplore.ieee.org/document/8667629
10. Lu, T.-J., et al.: Aluminum nitride integrated photonics platform for the ultraviolet to visible spectrum. Opt. Express **26**, 11147–11160 (2018)
11. Low-loss titanium dioxide waveguides and resonators using a dielectric lift-off fabrication process. https://opg.optica.org/oe/fulltext.cfm?uri=oe-23-9-11160&id=315568
12. Kaur, P. et al.: Hybrid and heterogeneous photonic integration. APL Photon. (2021)
13. Wunderer, T. et al.: Low-noise hybrid photonic integrated violet and blue lasers for quantum applications. In: 2022 IEEE Research and Applications of Photonics in Defense Conference (RAPID) 1–2 (IEEE, Miramar Beach, FL, 2022). https://doi.org/10.1109/RAPID54472.2022.9911592
14. Slight, T.J., Odedina, O., Meredith, W., Docherty, K.E., Kelly, A.E.: InGaN/GaN distributed feedback laser diodes with deeply etched sidewall gratings. IEEE Photon. Technol. Lett. **28**, 2886–2888 (2016)
15. Franken, C.A.A., et al.: Hybrid-integrated diode laser in the visible spectral range. Opt. Lett. **46**, 4904 (2021)
16. Menezo, S., et al.: Advances on III-V on silicon DBR and DFB lasers for WDM optical interconnects and associated heterogeneous integration 200mm-wafer-scale technology. In: 2014 IEEE Compound Semiconductor Integrated Circuit Symposium (CSICS), pp. 1–6 (IEEE, La Jolla, CA, 2014). https://doi.org/10.1109/CSICS.2014.6978538
17. Bovington, J.T., Heck, M.J.R., Bowers, J.E.: Heterogeneous lasers and coupling to Si_3N_4 near 1060 nm. Opt. Lett. **39**, 6017 (2014)
18. Xiang, C., et al.: Narrow-linewidth III-V/Si/Si_3N_4 laser using multilayer heterogeneous integration. Optica **7**, 20 (2020)
19. Tran, M.A., et al.: Extending the spectrum of fully integrated photonics to submicrometre wavelengths. Nature **610**, 54–60 (2022)
20. Kamei, T., et al.: Research toward a heterogeneously integrated InGaN laser on silicon. Phys. Status Solidi A **217**, 1900770 (2020)
21. Op De Beeck, C., et al.: Heterogeneous III-V on silicon nitride amplifiers and lasers via microtransfer printing. Optica **7**, 386 (2020)
22. Luo, X., et al.: High-throughput multiple dies-to-wafer bonding technology and III/V-on-Si hybrid lasers for heterogeneous integration of optoelectronic integrated circuits. Front. Mater. **2** (2015)
23. Li, W., et al.: Activation of buried p-GaN in MOCVD-regrown vertical structures. Appl. Phys. Lett. **113**, 062105 (2018)
24. Niklaus, F., Enoksson, P., Kalvesten, E., Stemme, G.: Void-free full wafer adhesive bonding. In: Proceedings IEEE Thirteenth Annual International Conference on Micro Electro Mechanical Systems (Cat. No.00CH36308), pp. 247–252 (2000). https://doi.org/10.1109/MEMSYS.2000.838524
25. Zhang, J.: High precision wafer thinning using ultra-low-TTV glass carrier and novel temporary bonding. In: 2023 IEEE 25th Electronics Packaging Technology Conference (EPTC), pp. 580–583 (2023). https://doi.org/10.1109/EPTC59621.2023.10457871
26. Ou-Yang, T.Y., Chang, H.H., Hsu, C.K.: Optimization of temporary wafer bonding materials and processes for 3D IC integration. In: 2020 15th International Microsystems, Packaging, Assembly and Circuits Technology Conference (IMPACT), pp. 36–39 (2020). https://doi.org/10.1109/IMPACT50485.2020.9268576
27. Montméat, P., et al.: Origin of the TTV of thin films obtained by temporary bonding ZoneBond® technology. Microsyst. Technol. **21**, 987–993 (2015)
28. Freestanding GaN-substrates and devices.pdf

29. Zhuang, D., Edgar, J.H.: Wet etching of GaN, AlN, and SiC: a review. Mater. Sci. Eng. R. Rep. **48**, 1–46 (2005)
30. Huaiyue, Y., et al.: Chemical mechanical polishing of freestanding GaN substrates. J. Semicond. **30**, 023003 (2009)

Chapter 12
All-Optical Activation Function with Sub-nanojoule Activation Energy Using VO_2-silicon Technology

Jorge Parra, Juan-Francisco Morcillo, Koen Schouteden, Jean-Pierre Locquet, and Pablo Sanchis

Abstract In this work, we report an all-optical nonlinear activation function with a triangular basis shape, implemented using vanadium dioxide (VO_2) integrated on a silicon-on-insulator (SOI) platform. The device exploits the insulator-to-metal transition (IMT) of VO_2 to achieve a nonlinear optical response with an ultra-compact footprint of just 2 μm. Experimental results demonstrate a low activation power of ~ 0.4 mW under continuous-wave operation and an activation energy of ~ 100 pJ with nanosecond response times under pulsed excitation. The proposed device operates across the telecom C-band with a high extinction ratio exceeding 10 dB, making it a promising candidate for large-scale neuromorphic photonic architectures.

12.1 Introduction

Artificial neural networks (ANNs), inspired by biological brains, have propelled breakthroughs in computational capabilities across diverse applications such as speech recognition, image classification, computer vision, and natural language processing [1]. The rapid expansion of ANN-based solutions has led to ever-increasing computational demands, straining electronic hardware as Moore's Law nears its limits. Photonic technologies offer a promising alternative to overcome these bottlenecks by exploiting the large bandwidth and low energy consumption of light [2]. In particular, integrated photonics has proven adept at linear operations—crucial for neural network interconnects—and can be readily implemented on silicon photonics platforms by leveraging mature complementary metal–oxide–semiconductor (CMOS)-compatible fabrication processes [3].

J. Parra · J.-F. Morcillo · P. Sanchis (✉)
Nanophotonics Technology Center, Universitat Politècnica de València, Valencia, Spain
e-mail: pabsanki@ntc.upv.es

K. Schouteden · J.-P. Locquet
Department of Physics and Astronomy, KU Leuven, Leuven, Belgium

P. Smowton and T. Krauss (eds.), *Proceedings of the 26th European Conference on Integrated Optics*, Springer Proceedings in Physics 440,
https://doi.org/10.1007/978-3-032-29673-3_12

A significant challenge for all-optical neural networks lies in realizing efficient nonlinear activation functions. Many proposed photonic activation solutions face trade-offs in footprint, bandwidth, and threshold power [4]. Vanadium dioxide (VO_2), a phase-change material featuring a reversible insulator-to-metal transition (IMT) under optical or thermal excitation, has emerged as a promising candidate to address these challenges. Its abrupt change in optical transmittance and electrical conductivity near room temperature (~65 °C) enables compact, low-energy photonic devices [5–9].

Here, we experimentally demonstrate an all-optical triangular basis-shaped activation function with sub-nanojoule activation energy, integrating VO_2 into a silicon waveguide structure. By harnessing the IMT of VO_2, our device achieves strong nonlinear responses over a broad spectral range while maintaining a small footprint and low activation energy.

12.2 Results

The working principle of the proposed activation function is sketched in Fig. 1a and b. For low optical powers, the optical response is linear with low loss as shown in Fig. 1b. By contrast, for certain optical powers, the nonlinear regime is activated, giving rise to a large increase in the optical loss of the device and thereby reducing the optical power at the output, as shown in Fig. 1b. Figure 1c shows an optical micrograph of the fabricated device. Our device consists of a 500 nm-wide and 220 nm-thick strip silicon waveguide patterned by e-beam lithography. A 40 nm-thick patch of VO_2 was grown by molecular beam epitaxy on top of the waveguide with a 2 μm diameter, forming the hybrid VO_2/Si section. The device was protected with a 1 μm-thick SiO_2 cladding deposited by plasma-enhanced vapor deposition. Fiber-to-chip coupling was achieved by standard grating couplers working for the transverse-magnetic (TM) polarization.

The input–output relationship of the device, in dB units, is modeled according to Eq. (12.1) [4], where P_{out} is the output power, P_{in} is the input power, P_{IMT} is the threshold power upon the IMT is triggered, P_{abs} is the absorbed power by the VO_2,

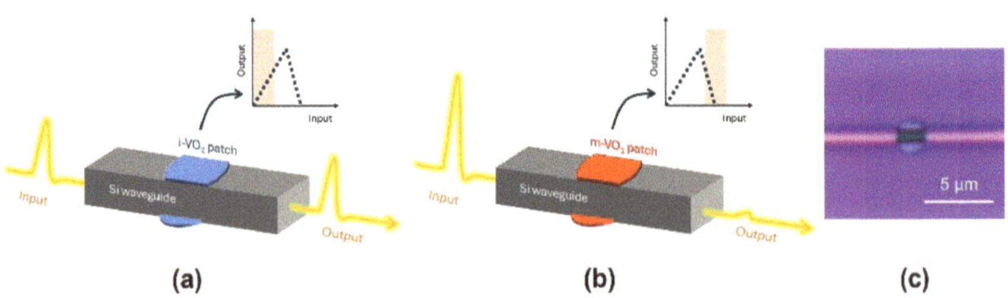

Fig. 12.1 Working principle of the proposed all-optical activation function device based on a VO_2/Si waveguide structure. **a** Linear regime in the insulating state (i-VO_2). **b** Nonlinear regime during the IMT until reaching the metallic state (m-VO_2). **c** Optical micrograph of the fabricated device

η is the thermo-optical coefficient relative to the rate conversion of the VO_2 length from insulator to metal, α_i and α_m are the propagation losses of the optical mode when the VO_2 is in the insulating and metallic states, respectively, L is the length of the device, $\Delta\alpha = \alpha_m - \alpha_i$, and $\Delta P = P_{abs} - P_{IMT}$.

$$P_{out} \approx \begin{cases} P_{in} - \alpha_i L, P_{abs} < P_{IMT} \\ P_{in} - \alpha_i L - \eta \Delta\alpha \Delta P, P_{abs} \geq P_{IMT} \end{cases} \quad (12.1)$$

To achieve a triangular-shaped optical response, the condition $\eta\Delta\alpha > 1$ must be fulfilled. According to our optical and thermal simulations, the TM fundamental mode at 1550 nm wavelength yields $\alpha_i = 2.6$ dB/μm, $\alpha_m = 7.5$ dB/μm, and $\eta \approx 0.5$ μm/dB, thereby satisfying the previous condition while featuring an activation power of $P_{IMT} \approx 0.5$ mW.

The measured optical performance of our activation function device is shown in Fig. 12.2. The spectral transmission response of the device in the insulating and metallic states is depicted in Fig. 2a. A Peltier device was employed to control the temperature of the photonic chip and, thereby, the phase state of the VO_2. The insertion loss of the device is below 3 dB with an extinction ratio exceeding 10 dB, which is in good agreement with our numerical simulations. Moreover, such values extend over the telecom C-band, showcasing the broadband response of the activation function. The experimental input–output relationship was measured at 1550 nm using a continuous-wave (CW) laser connected to an erbium-doped fiber amplifier (EDFA) prior to injecting the light into the chip. The nonlinear response is shown in Fig. 2b. A triangular-shaped response is achieved as predicted by our model with an activation power of ~ 0.4 mW, in agreement with thermal simulations.

On the other hand, the temporal dynamics of the device were also characterized using a pump-probe experiment. Sub-microsecond pump optical pulses at 1565 nm wavelength were generated using a fiber-based high-speed electro-optical modulator and an EDFA, whereas the probe was a low-power CW signal at 1550 nm wavelength. The pump-probe results are shown in Fig. 2c for different pump energies. The IMT

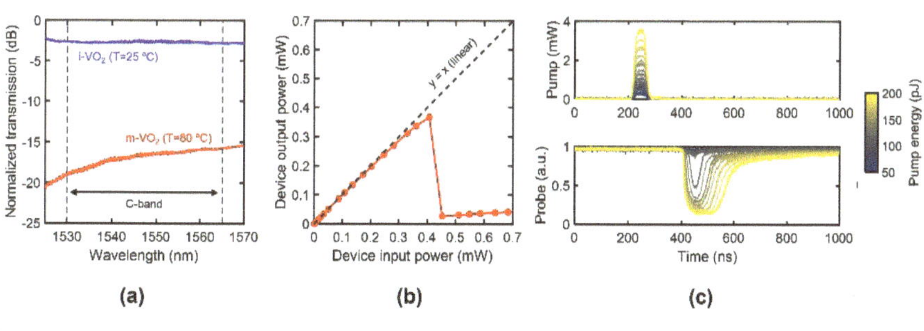

Fig. 12.2 Measured optical response of the VO_2/Si activation function device. **a** Optical spectra in the insulating and metallic states. **b** Experimental input–output relationship under continuous-wave signal. The dashed line stands for the linear response. **c** Temporal dynamics of the device using a pump-probe experiment, injecting different pulse energies into the device

is triggered for energy values ~ 100 pJ together with fall and rise times below 30 ns and 100 ns, respectively.

12.3 Conclusions

In this work, we have experimentally demonstrated an all-optical nonlinear device based on VO_2-Si technology for activation functionalities in photonic ANN. Our device features an ultracompact footprint of 2 μm together with broad operation across the telecom C-band, low activation power of just ~ 0.4 mW under CW operation, and low activation energies of ~ 100 pJ with nanosecond speed under pulse excitation. Our results highlight the potential of VO_2-Si technology for enabling large-scale photonic ANN architectures such as deep neural networks based on wavelength-division multiplexing.

Acknowledgements This work is supported by the following funding: Agencia Estatal de Investigación (TED2021-132211B-I00); European Health and Digital Executive Agency (101070690 (PHOENIX)); Agencia Estatal de Investigación (PID2022-137787OB-I00); Universitat Politècnica de València (Grants PAID-06-23, PAID-10-23). The authors would like to thank the clean room staff of the Nanophotonics Technology Center for their help with the fabrication of the chip.

References

1. Goldberg, Y.: A primer on neural network models for natural language processing. J. Artif. Intell. Res. **57**, 345–420 (2016)
2. Shastri, B.J., et al.: Photonics for artificial intelligence and neuromorphic computing. Nat. Photon. **15**, 102–114 (2021)
3. El Srouji, L., et al.: Photonic and optoelectronic neuromorphic computing. APL Photon. **7**, 051101 (2022)
4. Parra, J., Navarro-Arenas, J., Sanchis, P.: Sub-milliwatt threshold power and tunable-bias all-optical nonlinear activation function using vanadium dioxide for wavelength-division multiplexing photonic neural networks. Sci. Rep. **15**, 5608 (2025)
5. Cueff, S., et al.: VO_2 nanophotonics. APL Photonics **5**, 110901 (2020)
6. Wong, H.M.K., et al.: Broadband, integrated, micron-scale, all-optical Si_3N_4/VO_2 modulators with pJ switching energy. ACS Photon. **6**, 2734–2740 (2019)
7. Parra, J., et al.: Low-threshold power and tunable integrated optical limiter based on an ultracompact VO_2/Si waveguide. APL Photon. **6**, 121301 (2021)
8. Jung, Y., et al.: Integrated hybrid VO_2–silicon optical memory. ACS Photon. **9**, 217–223 (2022)
9. Seoane, J.J., et al.: Ultra-high endurance silicon photonic memory using vanadium dioxide. NPJ Nanophoton. **1**, 37 (2024)

Chapter 13
Towards Laser-Based Optic Manufacturing for Novel Fiber-Chip-Couplers through Selective Laser-Induced Etching

Sandra Borzek, Jana Köller, Georgios Sinatkas, Manuel Jung, Anton Schillings, and Jonathan Müller

Abstract Selective Laser-induced Etching (SLE) is a two-stage process that enables the manufacturing of geometry-free and highly precise monolithic 3D components. First, the shape of the component is imprinted into the glass using tightly focused, ultrashort pulsed laser radiation. In a second step, wet chemical etching is applied to expose the component from the surrounding material. As the miniaturization of microelectronics and photonic integrated circuits (PICs) approaches its limits with conventional techniques, the SLE process offers a promising solution for fabricating innovative fiber-chip-couplers for PIC-based external cavity diode lasers in quantum technology applications. These fiber-chip-couplers demand extremely low alignment tolerances, high precision, minimal coupling loss, and resilience to external thermal and atmospheric factors. A crucial aspect of achieving these requirements is the precise positioning of the integrated optics in the micrometer regime. In the following, we will therefore examine the laser-based fabrication of microlenses for potential monolithic or semi-monolithic integration into the fiber-chip-coupler. We will present a process chain that combines the SLE process with laser polishing, and subsequently analyze the microlenses in terms of the achieved micro-roughness and their lens profile. Furthermore, we will present strategies on how such optics within the overall component can be manufactured.

S. Borzek (✉) · J. Köller · M. Jung · A. Schillings · J. Müller
Fraunhofer Institute for Laser Technology ILT, Aachen, Germany
e-mail: sandra.borzek@ilt.fraunhofer.de

G. Sinatkas
Institute for Integrated Photonics IPH, RWTH Aachen, Aachen, Germany

P. Smowton and T. Krauss (eds.), *Proceedings of the 26th European Conference on Integrated Optics*, Springer Proceedings in Physics 440,
https://doi.org/10.1007/978-3-032-29673-3_13

13.1 Introduction

Small and robust external cavity diode lasers (ECDLs) based on photonic-integrated circuits (PICs) play a crucial role in various quantum technology applications. Relevant wavelengths for quantum applications include 637 nm for nitrogen vacancy centers, which serve as excellent single-photon emitters [1, 2]; 650 nm for the barium quantum computer [3]; 397 nm for the calcium quantum computer [3]; and 461, 689, 679, and 707 nm for the strontium atomic clock [4]. Current high-end laser systems for such applications are quite large and unsuitable for use outside of laboratories. New and customized designs for the miniaturization of quantum hardware to facilitate the integration of quantum technologies into everyday life are enabled by Selective Laser-induced Etching (SLE). This study presents the current development of the laser-based fabrication of micro-optics, aiming to integrate SLE-fabricated optics directly into a novel fiber-chip-coupler.

SLE is a two-stage method for processing transparent dielectrics that can be used for the fabrication of high-precision, geometry-free 3D components. Initially, a 3D CAD model is created, which is then converted into modification vectors through semi-automatic modeling. These vectors are positioned in planes perpendicular to the irradiation direction of the laser. In the subsequent laser processing step, ultrashort pulsed (USP) laser radiation is tightly focused into the material, achieving intensities exceeding 10^{13} W/cm^2. This surpasses the ionization thresholds of transparent dielectrics through nonlinear absorption mechanisms at visible and near-visible wavelengths, resulting in local structural changes and inscribing the modification vectors into the material. By moving the laser relative to the dielectric, contiguous areas within the volume and on the surface can be modified. In the following step, the processed sample is etched in an ultrasonic bath with a suitable etching agent. The structured areas show a significantly higher etching rate with wet etching agents compared to the unexposed material. The ratio of the etch rate of the modified material R_{mod} [μm/h] to that of the unmodified material R_{unmod} [μm/h] is known as selectivity S

$$S = 1 + \frac{R_{mod}}{R_{unmod}} \tag{13.1}$$

and quantifies the SLE process. Notably, selectivities of up to 1400:1 can be achieved with KOH in fused silica, underscoring its suitability for producing monolithic 3D glass components in the micrometer range. The structural dimensions of these 3D components can vary from centimeters to micrometers [5]. The applications of the SLE process are diverse, spanning microfluidics, medical [6–11] and sensor industries [12–14], optical fabrication [15–17], and quantum hardware [18–20]. An illustration of the process is shown in Fig. 13.1.

The component surfaces produced by SLE exhibit a roughness in the micrometer range, requiring a subsequent polishing step to produce surfaces with optical quality. Laser polishing using a CO_2 laser beam source is studied, during which the glass

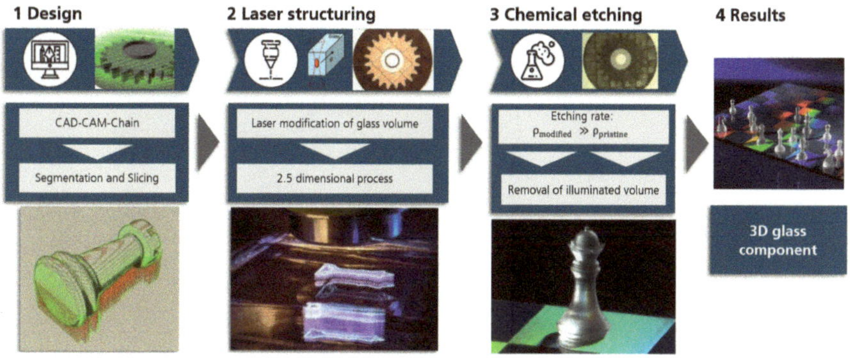

Fig. 13.1 Process chain of the SLE process [21]

surface is heated by a quasi-line of the laser beam. Smoothing takes place without significant material removal due to surface tension at reduced viscosity. This enables roughness peaks to flow into valleys by material redistribution, effectively reducing overall roughness down to Sa < 1 nm [22, 23]. However, the surface shape cannot be precisely controlled.

13.2 Materials and Methods

13.2.1 Experimental Setup

The USP beam source used to induce modifications within the bulk material is an Amplitude Satsuma HP (Amplitude Systems) with a wavelength of $\lambda = 1030$ nm and an average power of $\overline{P} = 10$ W. The pulse repetition rate can be varied from $f_{Rep} = 1$ Hz to 40 MHz, and the pulse duration range is from $\tau = 380$ ps to 10 ps. The raw beam diameter is expanded to 6 mm using a telescope (*OPTOGAMA MEXI8*) to fully illuminate the focusing optics' aperture. The linearly polarized laser beam is transformed into circular polarization by a beam splitter in combination with a lambda quarter plate. The subsequent microscanner system comprises a galvanometer scanner (*SCANLAB intelliSCAN$_{se}$14*) and a focusing optic (*Mitutoyo M Plan Apo NIR 20x/0.4*) with NA = 0.4. These optical components and the scanner system are mounted on a linear axis (*Aerotech ATS50030*). The sample holder features a vacuum chuck (*LightFab*), a tilting stage, and a rotation stage (*Edmund Optics*) for micrometer-precise sample alignment. It is mounted on an air-bearing axis system (*Aerotech ABL10100-LN and Aerotech ABL20030*). The maximum size of the scan field covered by the scanner is 500 μm × 500 μm, limited by the focusing optics. To structure larger components, multiple scan fields are stitched together. A schematic diagram of the utilized setup is presented in Fig. 13.2.

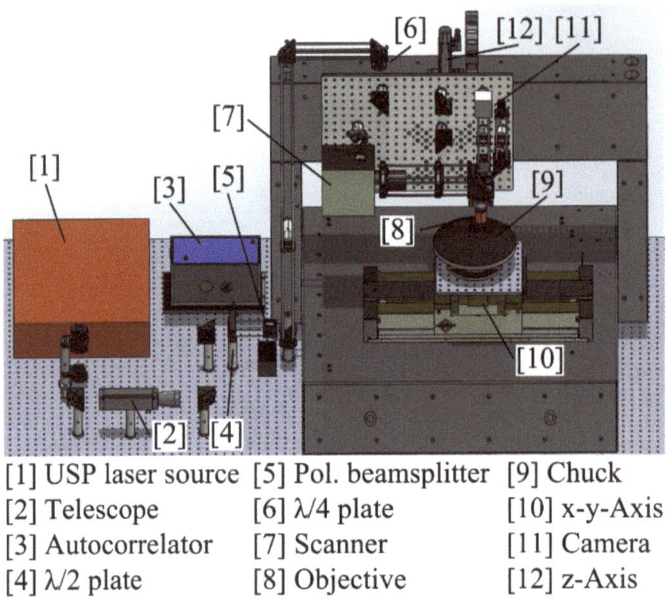

Fig. 13.2 Schematic drawing of the system layout [24]

The manufactured lenses are analyzed using a *Keyence VHX 6000* digital microscope, a *Zygo NewView™ 9000* white light interferometer (WLI) and a *Keyence VK-X3000* laser scanning microscope (LSM).

13.2.2 Model Design and Fabrication

The fiber-chip-coupler measures 28.50 × 10 × 10 mm^3. The laser beam is generated in a PIC and is guided directly through a waveguide into the fiber-chip-coupler. The beam is then collimated, passes through an optical isolator and a beam splitter, and is finally focused into a single-mode fiber (SM fiber) (cf. Fig. 13.3).

As a preliminary step in integrating optics into the fiber-chip-coupler, a plano-convex lens is fabricated on a carrier with protruding edges. This carrier enables precise analysis of the lens's centricity and allows for testing the effects of laser polishing on surrounding structures (cf. Fig. 13.3).

There are two primary methods for integrating SLE-manufactured optics: Regarding the goal of monolithic design, the lenses can be integrated upright into the housing of the fiber-chip-coupler. This approach is challenging for the laser polishing step, as the CO_2 laser beam cannot access the surface of the lenses through the component's housing. One solution is to fabricate the fiber-chip-coupler in two parts and to bond them seamlessly.

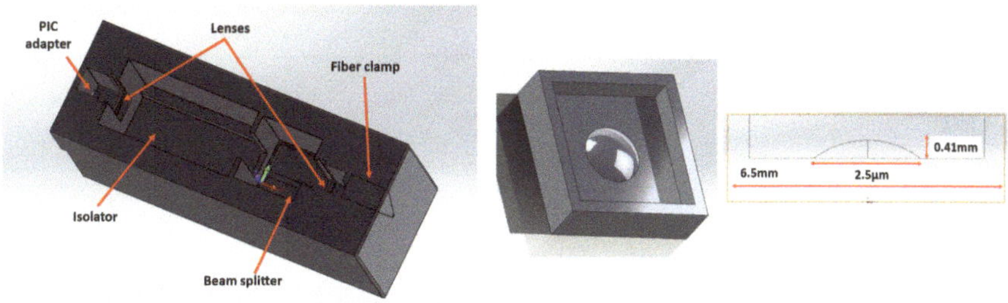

Fig. 13.3 Left: Model of the fiber-chip-coupler, including spaces for an optical isolator, a beam splitter, a fiber clamp, lens insertion devices, and a PIC adapter. Right: Model of the lens mounted on carrier, featuring a diameter of 2.5 mm and a center thickness of 0.41 mm

Fig. 13.4 Model of a lens on a slide-in carrier

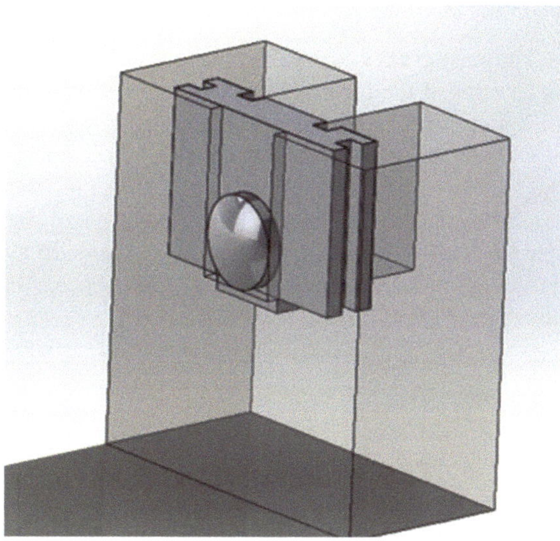

Alternatively, the lenses can be produced separately on a carrier and inserted into the fiber-chip-coupler without gaps, utilizing individual slide-in structures (cf. Fig. 13.4). After insertion, the optics can be joined by laser polishing to the fiber-chip-coupler's surface, effectively fusing the glass material.

13.2.3 Material and Parameters

The fiber-chip-coupler and the lens are made from double-sided polished fused silica wafers (Siegert Wafer GmbH) with a thickness of 10 mm and 2.3 mm, respectively. The process parameters used to manufacture the lens are listed in Table 13.1.

Table 13.1 Process parameters to manufacture the lens on a carrier

Parameter	Value
Wavelength	1030 nm
Pulse duration	1.0 ps
Repetition rate	760 kHz
Scan velocity	200 mm/s
Pulse energy	293–533 nJ
Pulse energy correction	104 nJ/mm
Polarization	Circular
Etching fluid	KOH
Etching fluid concentration	8 Mol/L
Etching temperature	80 °C

The pulse energy correction compensates for changes of the local intensity caused by changes of the focal volume due to spherical aberration at greater depths of the glass material by adjusting the pulse energy.

Despite its significant advantages, such as geometric freedom, SLE has limitations that must be considered during the process. For instance, stress formation during laser processing can lead to cracks, influenced by energy deposition and the geometry and density of modifications. The spacing between modification vectors is crucial for resolution; however, too closely spaced vectors can increase stress and cracking.

13.3 Results

The duration of the laser-modification process is about 20 minutes, followed by 24 h of etching in potassium hydroxide (KOH). In Fig. 13.5 the centricity of the lens on the carrier is illustrated. The lens position was measured relative to the four corners of the substrate. A centricity and positional accuracy of ± 1 μm is achieved, demonstrating the capability of precise positioning using SLE.

No damage or cracks are visible on the surface of the lens, but grid-like structures from the laser scan field boundaries are clearly visible. The focus across the scan field in z-direction is not uniform due to beam propagation and deflection during scanning. The average intensity of the laser beam decreases towards the edges, potentially creating a height difference when stitching multiple scan fields.

To assess the impact of these scan field edges on the lens surface quality after laser polishing, micro-roughness measurements are taken before and after polishing using a WLI. The best-fit sphere is subtracted from the measurement results to isolate the pure micro-roughness excluding the curvature of the lens. Additionally, a high-pass filter with a cut-off wavelength of $\lambda = 10$ μm is applied, in accordance with [25] as definition of micro-roughness. This results in a surface micro-roughness of $Sa_{micro} = 4.35$ μm for the lens after SLE. The lens is then polished using a CO_2

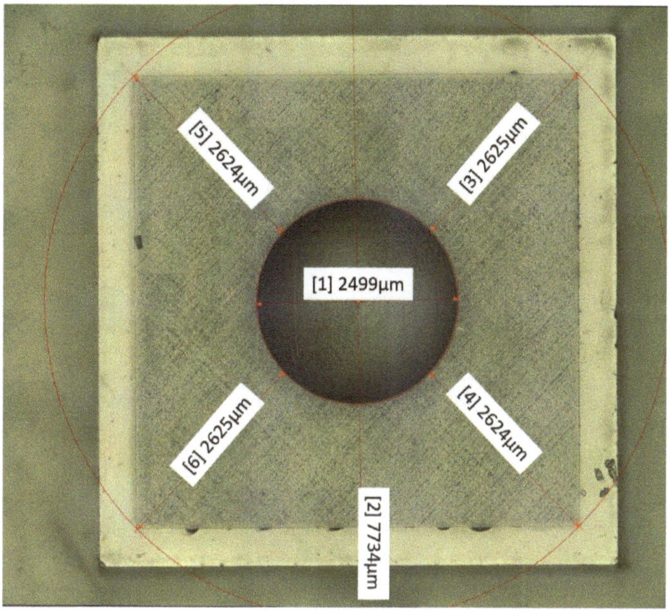

Fig. 13.5 Depth-of-field panoramic microscope image of the lens along with measurements of the lens centering. Auxiliary geometries include a circle drawn around the lens and circles at the four corners of the carrier

laser at an average power of $\overline{P} = 75$ W and a scanning speed of $v_s = 750$ mm/s. In the three-dimensional image captured with the WLI, no residues of the scan field edges are visible within the scope of resolution. However, after subtracting the best-fit sphere, artifacts contributing to shape deviation appear (cf. Fig. 13.6), with the micro-roughness measured to be $Sa_{micro} = 90.97$ nm. This yields a 97.90% reduction of the SLE micro-roughness.

To eliminate artifacts from the scan fields, a second lens is structured using a smaller scan field of 200 µm × 200 µm (test series 2). This smaller scan field reduces height differences between adjacent areas. While the structuring time for the lens increases from approximately 20 min to nearly an hour, the etching time remains unchanged. The scan field edges are still visible in the lens structured with a 200 µm scan field, but they are less pronounced (cf. Fig. 13.7).

The micro-roughness of the SLE blank $Sa_{micro} = 210.25$ nm is significantly reduced compared to $Sa_{micro} = 4.35$ µm generated by the 500 µm scan field. Following the initial laser polishing, it was observed that the SLE lenses from test series 2 experienced greater evaporation than those from test series 1, resulting in fused silica residues on the lens surface. These residues were re-fused during a second polishing, conducted at a reduced laser power of $\overline{P} = 65$ W and an increased scanning speed of $v_s = 5000$ mm/s, after the lens was fully cooled down. The resulting micro-roughness is $Sa_{micro} = 2.68$ nm, which is 1.27% of the initial micro-roughness of the SLE

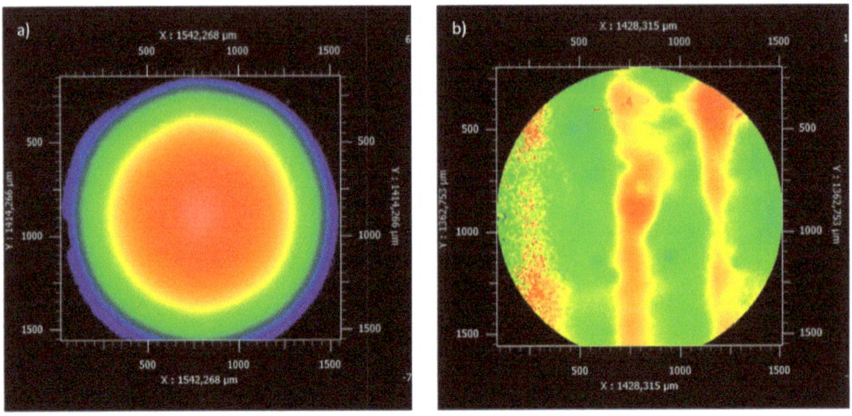

Fig. 13.6 False-color height images (magnification 5.5×) of **a** the total lens and **b** after subtraction of the best-fit sphere with a scan field size of 500 µm × 500 µm, revealing the artifacts of the scan field boundaries

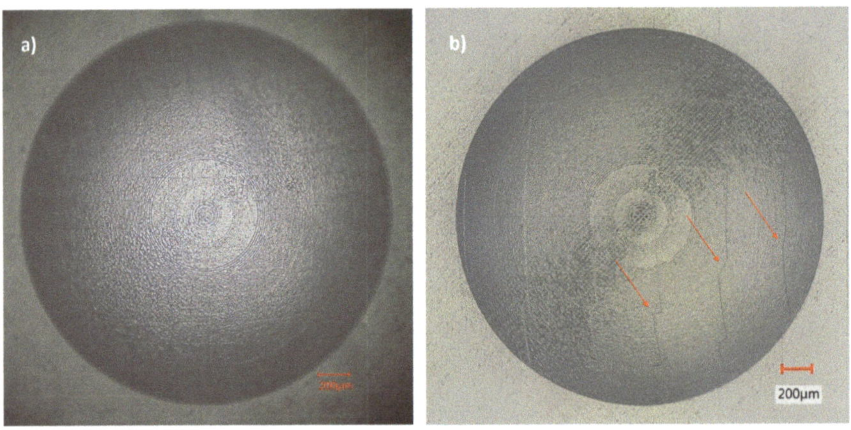

Fig. 13.7 Microscope images of lenses structured with **a** 200 µm × 200 µm scan field and **b** 500 µm × 500 µm scan field with red arrows pointing to the scan field boundaries. In (**a**), the scan field boundaries are significantly less pronounced

blank and represents an additional reduction of approximately 97% compared to the polished SLE lens from test series 1 (cf. Fig. 13.8).

To compare the profiles of the lens after SLE and after polishing with the initial model, surface profiles were measured using a LSM. The raw data and numerical fits are shown in Fig. 13.9.

The SLE blank shows a slight deviation of the lens center thickness compared to the model, being marginally flatter at the apex. This can be attributed to the scanning strategy, as the flat curvature at the apex becomes unresolved by the scan vectors,

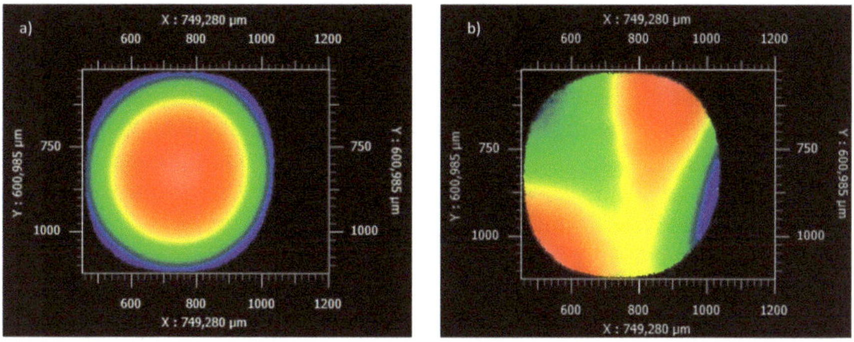

Fig. 13.8 False-color height images (magnification 5.5×) of **a** the total lens and **b** after subtraction of the best-fit sphere with a scan field size of 200 µm × 200 µm revealing remaining surface waviness

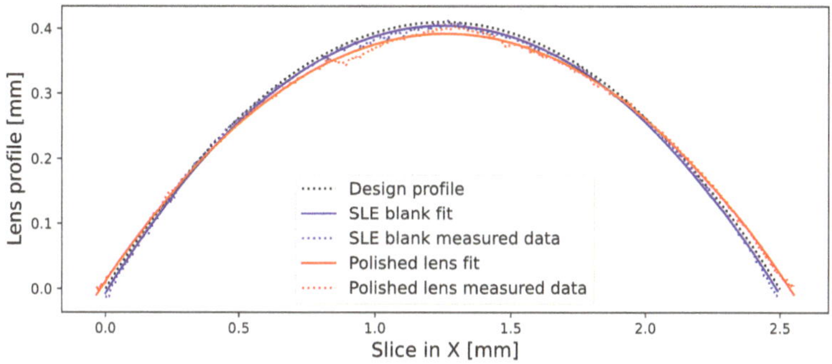

Fig. 13.9 Plot of the SLE lens profiles before and after laser polishing (blue and red respectively) and the modelled lens (black). The polished lens is slightly flatter with a less steep surface profile

leading to the formation of a plateau. Height measurements indicate an average center thickness of 408 µm deviating by 2 µm from the target.

After laser polishing, excess material is smoothed and redistributed to the edges, further flattening the lens, and reducing the flank gradient. A height difference of 7 µm at the lens apex was observed between the SLE blank and the polished lens. A second polishing run may increase the deviation from the target geometry. We believe that residues necessitating this second polish arise from the evaporation of fused silica in the first polishing steps, particularly at the carrier edges, indicating that surrounding protruding material affects the polishing of lenses. Additionally, stress induced during polishing can lead to deformation of the lens profile. If changes of the shape occur reproducibly during polishing, adjustments to the initial SLE model can compensate for deviations from the target geometry.

13.4 Conclusion

This study reviews the current status and potential for manufacturing optics for novel fiber-chip-couplers made of fused silica for highly integrated PIC-based ECDLs using SLE. Limitations regarding surface roughness for optical production can be addressed by subsequent laser polishing with a CO_2 laser. We have demonstrated that the microroughness of SLE micro lenses with a diameter of 2.5 mm can be reduced to 2.68 nm by the polishing process. Current investigations aim to adapt the initial SLE model to achieve the target lens geometry after laser polishing. The research and application of quantum technologies is a rapidly evolving field requiring highly specialized hardware. The SLE process serves as an effective tool for producing flexible and precise components as well as customized optics. To meet the demands for continuously evolving components, the SLE process needs to be further advanced in the future, both as a standalone method and in combination with other process steps like laser polishing.

Acknowledgements We acknowledge funding by the German ministry of science and education (BMBF) under grant 13N15966.

References

1. Acosta, V., Hemmer, P.: Nitrogen-vacancy centers: Physics and applications. MRS Bull. **38**(2), 127–130 (2013). https://doi.org/10.1557/mrs.2013.18
2. Jezelzko, F., Neumann, P., Rempp, F., Schnitzler, W., Singer, K., Schmidt-Kaler, F., Wrachtrup, J.: Ein Quantencomputer in Diamant, vol. 5. Universität Stuttgart (2008). https://www.uni-stuttgart.de/presse/archiv/themenheft/05/ein_quantencomputer_in_diamant.pdf. Accessed 26 Feb 22025
3. Dietrich, M.R., Avril, A., Bowler, R., Kurz, N., Salacka, J.S., Shu, G., Blinov, B.B., Danielson, J.R., Pedersen, T.S.: Barium Ions for quantum computation. In: AIP Conference Proceedings. Non-neutral Plasma Physics VII: Workshop on Non-neutral Plasmas, New York (New York), 16–20 June 2008, pp. 25–30. AIP (2009). https://doi.org/10.1063/1.3122286
4. Bober, M., Zachorowski, J., Gawlik, W., Morzyński, P., Zawada, M., Lisak, D., Cygan, A., Bielska, K., Piwiński, M., Trawiński, R.S., Ciuryło, R., Ozimek, F., Radzewicz, C.: Precision spectroscopy of cold strontium atoms, towards optical atomic clock. Bull. Polish Acad. Sci. Tech. Sci. **60**(4), 707–710 (2012). https://doi.org/10.2478/v10175-012-0082-x
5. Gottmann, J., Hermans, M., Repiev, N., Ortmann, J.: Selective laser-induced etching of 3D precision quartz glass components for microfluidic applications—up-scaling of complexity and speed. Micromachines. **8**(4), 110 (2017). https://doi.org/10.3390/mi8040110
6. Osellame, R., Hoekstra, H., Cerullo, G., Pollnau, M.: Femtosecond laser microstructuring: an enabling tool for optofluidic lab-on-chips. Laser Photon. Rev. **5**(3), 442–463 (2011). https://doi.org/10.1002/lpor.201000031
7. Osellame, R., Maselli, V., Vazquez, R.M., Ramponi, R., Cerullo, G.: Integration of optical waveguides and microfluidic channels both fabricated by femtosecond laser irradiation. Appl. Phys. Lett. **90**(23) (2007). https://doi.org/10.1063/1.2747194
8. Sugioka, K., Cheng, Y.: Femtosecond laser processing for optofluidic fabrication. Lab Chip. **12**(19), 3576–3589 (2012). https://doi.org/10.1039/c2lc40366h

9. Vitali, V., Nava, G., Zanchetta, G., Bragheri, F., Crespi, A., Osellame, R., Bellini, T., Cristiani, I., Minzioni, P.: Integrated optofluidic chip for oscillatory microrheology. Sci. Rep. **10**(1), 5831 (2020). https://doi.org/10.1038/s41598-020-62628-1
10. Kim, S., Kim, J., Joung, Y.-H., Ahn, S., Choi, J., Koo, C.: Optimization of selective laser-induced etching (SLE) for fabrication of 3D glass microfluidic device with multi-layer micro channels. Micro Nano Syst. Lett. **7**(1) (2019). https://doi.org/10.1186/s40486-019-0094-5
11. Yu, J., Xu, J., Zhang, A., Song, Y., Qi, J., Dong, Q., Chen, J., Liu, Z., Chen, W., Cheng, Y.: Manufacture of three-dimensional optofluidic spot-size converters in fused silica using hybrid laser microfabrication. Sensors (Basel, Switzerland) **22**(23) (2022). https://doi.org/10.3390/s22239449
12. Song, J., Lin, J., Tang, J., Liao, Y., He, F., Wang, Z., Qiao, L., Sugioka, K., Cheng, Y.: Fabrication of an integrated high-quality-factor (high-Q) optofluidic sensor by femtosecond laser micromachining. Opt. Express. **22**(12), 14792–14802 (2014). https://doi.org/10.1364/OE.22.014792
13. Beckmann, L., Bi, T., Thoms, J.M., Wenk, M., Zhang, S., Kratz, M., Del'Haye, P.: Selective laser-induced etching for novel 3D microphotonic devices. In: Kling, R., Pfleging, W., Sugioka, K. (eds.) Laser-based Micro- and Nanoprocessing XVIII. Laser-based Micro- and Nanoprocessing XVIII, San Francisco, United States, 27 Jan 2024–01 Feb 2024, p. 3. SPIE (2024). https://doi.org/10.1117/12.3002335
14. He, F., Liao, Y., Lin, J., Song, J., Qiao, L., Cheng, Y., Sugioka, K.: Femtosecond laser fabrication of monolithically integrated microfluidic sensors in glass. Sensors (Basel, Switzerland). **14**(10), 19402–19440 (2014). https://doi.org/10.3390/s141019402
15. McMillen, B., Athanasiou, C., Bellouard, Y.: Femtosecond laser direct-write waveplates based on stress-induced birefringence. Opt. Express. **24**(24), 27239–27252 (2016). https://doi.org/10.1364/OE.24.027239
16. Ross, C., MacLachlan, D.G., Choudhury, D., Thomson, R.R.: Towards optical quality micro-optic fabrication by direct laser writing and chemical etching. In: Heisterkamp, A., Herman, P.R., Meunier, M., Osellame, R. (eds.) Frontiers in Ultrafast Optics: Biomedical, Scientific, and Industrial Applications XVII. SPIE LASE, San Francisco, California, United States, Saturday 28 Jan 2017, 100940V. SPIE (2017). https://doi.org/10.1117/12.2251986
17. Skora, J.-L., Gaiffe, O., Bargiel, S., Cote, J.-M., Tavernier, L., de Labachelerie, M., Passilly, N.: High-fidelity glass micro-axicons fabricated by laser-assisted wet etching (2021). https://doi.org/10.48550/arXiv.2110.12212
18. Simeth, S., Müller, A., Müller, J., Lekitsch, B., Reininghaus, M., Schmidt-Kaler, F.: Selective laser-induced etching for 3D ion traps. In: Kling, R., Pfleging, W., Watanabe, A. (eds.) Laser-based Micro- and Nanoprocessing XVII. Laser-based Micro- and Nanoprocessing XVII, San Francisco, United States, 28 Jan 2023–03 Feb 2023, p. 2. SPIE (2023). https://doi.org/10.1117/12.2647189
19. Peters, C., Kratz, M., Köller, J., Borzek, S., Simeth, S.: Selective laser-induced etching (SLE) of transparent materials for microelectronic components and quantum computing applications. In: Kling, R., Pfleging, W., Sugioka, K. (eds.) Laser-based Micro- and Nanoprocessing XVIII. Laser-based Micro- and Nanoprocessing XVIII, San Francisco, United States, 27 Jan 2024–01 Feb 2024, p. 20. SPIE (2024). https://doi.org/10.1117/12.2692475
20. Corrielli, G., Crespi, A., Osellame, R.: Femtosecond laser micromachining for integrated quantum photonics (2021). https://doi.org/10.48550/arXiv.2110.06162
21. Kratz, M., Beckmann, L., Borzek, S., Peters, C., Saßmannshausen, A.: Manufacturing optical components by selective laser-induced etching (conference contribution). In: LaP (6th Conference on Laser Polishing)/LOM (1st Conference on Laser-Based Optics Manufacturing), Aachen, Germany, vol. 16, pp. 10 (2024)
22. Hecht, K.: Entwicklung eines Laserstrahlpolierverfahrens für Quarzglasoberflächen. Dissertation, Technische Universität Ilmenau (2012)
23. Jung, M.: Impact of preheating conditions on form deviation during laser polishing of N-BK7 glass. In: Thieß, H., Fütterer, G., Wünsche, C., Fähnle, O.W., Haberl, A. (eds.) Eleventh European Seminar on Precision Optics Manufacturing. Eleventh European Seminar on Precision

Optics Manufacturing, Teisnach, Germany, 09 Apr 2024–10 Apr 2024, p. 1. SPIE (2024). https://doi.org/10.1117/12.3030903
24. Peters, C., Geppert, B., Dashjav, E., Kratz, M., Tietz, F.: Glass molding tools fabricated with selective laser-induced etching for sustainable treatment of solid-state electrolytes. JLMN (2024). https://doi.org/10.2961/jlmn.2024.01.2014
25. Temmler, A.: Selektives Laserpolieren von metallischen Funktions- und Designoberflächen. Dissertation, Rheinisch-Westfälische Technische Hochschule Aachen (RWTH) (2013)

Chapter 14
Homodyne CW-THz System Driven by Photonic Integrated Beat Signal Source

Niklas Schulz, Nils Surkamp, Lisa C. Kreuzer, Carsten Brenner, and Martin R. Hofmann

Abstract This work examines photonic integrated lasers for their applicability as a beat signal source for homodyne CW-THz systems. Two DFB lasers on a custom designed PIC are analyzed for their coarse and fine tunability in wavelength. Employed in a CW-THz system the fine tunability of the integrated lasers enables phase sampling without additional components. This is demonstrated by transmissive thickness measurements of ABS samples.

14.1 Introduction

Terahertz (THz) technology, operating in the electromagnetic spectrum between 100 GHz and 10 THz, offers many versatile properties. Especially its ability to penetrate non-conductive materials, provide high-resolution imaging, and offer precise spectroscopic identification of chemical compounds, all while being non-ionizing and safe for human exposure, enable a wide range of applications across various fields, that include non-destructive testing, security screening, and high-speed wireless communication [1, 2].

Photonics based continuous wave (CW) THz systems rely on the superposition of two optical modes with different optical frequencies to create a beat signal. The difference frequency part of this signal is radiated as a continuous THz wave by means of a photodiode-based emitter antenna. In a homodyne system scheme, the THz wave is mixed with the same optical beat signal at the photoconductive receiver antenna. In this mixing process, the relative phase $\Delta\varphi$ of the incident THz wave to the optical beat signal at the receiver is important. The photocurrent generated at the receiver can be described by

$$\mathrm{I_{det}} \propto \mathrm{E_{THz}} \cos(\Delta\varphi), \quad \Delta\varphi = \frac{2\pi \cdot \mathrm{f_{THz}} \cdot \mathrm{L}}{\mathrm{c}} + \varphi_{\mathrm{mod}} \tag{14.1}$$

N. Schulz (✉) · N. Surkamp · L. C. Kreuzer · C. Brenner · M. R. Hofmann
Photonics and Terahertz Technology, Ruhr University Bochum, Bochum, Germany
e-mail: niklas.schulz@ruhr-uni-bochum.de

P. Smowton and T. Krauss (eds.), *Proceedings of the 26th European Conference on Integrated Optics*, Springer Proceedings in Physics 440,
https://doi.org/10.1007/978-3-032-29673-3_14

where E_{THz} is the electrical field strength of the THz wave, f_{THz} is its frequency and L is the optical length difference between receiver arm and the transmitter arm including the THz path. To reconstruct the THz amplitude and phase information at the desired frequency point, $\Delta\varphi$ needs to be sampled. For this, mainly three different approaches are available. First, the phase of the beat signal can be shifted by a mechanical free space delay line in either arm changing the relative path length L. The relative phase of the beat signals can also be changed by introducing a mode selective phase variation in the optical domain (φ_{mod}). The third approach leverages the influence of f_{THz} on the relative phase. With a non-zero path length difference L, a change in THz frequency also results in the change of $\Delta\varphi$. While this approach requires fine tunability of the optical modes, it doesn't require additional components in the setup [3, 4].

Especially at a wavelength around 1.5 μm compact and stable continuous wave lasers as well as mature optical fiber technology are available as a backbone for photonics based CW-THz systems. Photonic integration, which involves the consolidation of multiple photonic functions onto a single chip, offers even further advantages in terms of size, weight, power consumption, and cost and potentially paves the way for mass deployment of THz systems.

This paper tests two superimposed distributed feedback (DFB) lasers on a custom designed photonic integrated circuit (PIC) as a beat signal source for a homodyne CW-THz system as a first step towards fully integrated THz systems.

14.2 Experimental

The PIC is designed with Luceda's IPKISS design platform and fabricated by Fraunhofer HHIs multi project wafer service. Besides the two DFB lasers which are tested in this work, it incorporates a sampled grating distributed feedback laser (SGDBR) as well as semiconductor optical amplifiers (SOA) and thermo-optic phase sections (TOP) to investigate multiple arrangements of the lasers and phase sampling of the THz signal (s. Fig. 14.1a). In this work, the tunability of both DFB lasers is investigated for applicability in a homodyne CW-THz system utilizing a controlled frequency sweep for phase sampling. For this, the two DFB modes are superimposed on chip by a multimode interferometer (MMI) and guided to a spot size converter (SSC) to couple the beat signal in an optical fiber. The DFB modes can be tuned in emission frequency via the gain current and via current driven heater elements placed on top of the active area of each laser. The PIC has a size of 2 mm by 8 mm and is mounted on a temperature controlled base. The DFB gain and heater segments are connected to individual current sources.

First the optical emission from both DFBs is measured at different gain currents with an optical spectrum analyzer (*Yokogawa AQ6370B*). Both DFBs are tuned from 0 to 120 mA and the recorded spectra are shown in Fig. 14.1b exhibiting a red-shift for higher currents. Each DFB can be tuned by about 4 nm (500 GHz) via the gain current, resulting in a continuous difference frequency range of 200 GHz to 1.2 THz.

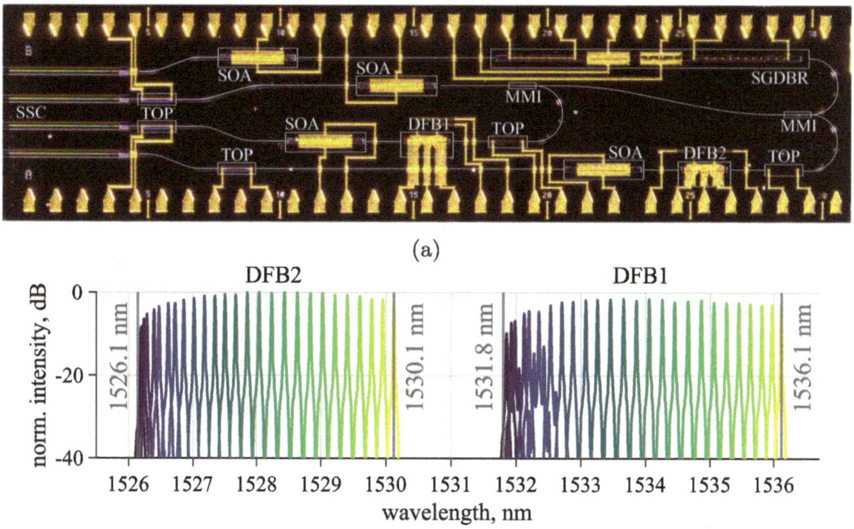

(b)

Fig. 14.1 **a** Annotated microscope image of the PIC. Components are surrounded by white boxes and labeled. Waveguides are highlighted by white lines. **b** Emission spectra of both DFB lasers for gain currents of 20 mA (dark) to 120 mA (bright) coupled from the second SSC from top

The heater elements on the gain segments can red-shift the laser's emission by an additional 0.2 nm for a heater current of 0–30 mA. The 3dB loss in power of DFB1, compared to DFB2, is attributed to the additional MMI in the signal path of DFB1. The lasers are spaced about 2.5 mm apart leading to minor thermal crosstalk. While DFB1 is tuned by 4 nm and the gain current of DFB2 is held constant, DFB2's emission is shifting about 20 pm, which is acceptable for the intended application.

The PIC is now employed in a homodyne CW-THz system (s. Fig. 14.2). The beat signal from the PIC (about 4 mW) needs to be boosted by an Erbium-doped fiber amplifier (EDFA, *Thorlabs EDFA100P*) before being divided to meet the power requirements of the system (30 mW per antenna). The optical beat signal at the transmitter along with a -1.4 V bias voltage creates the CW-THz signal which is guided through the setup and focussed on the sample in transmission geometry using four off-axis parabolic mirrors (1$''$ diameter, 2$''$ focal length). The photocurrent generated at the receiver antenna, described by (14.1), is converted to a voltage by a transimpedance amplifier (TIA, *Femto DHPCA-100*) which is acquired by direct sampling (DAQ, *Zurich Instruments UHFLI*). The receiver arm contains a mechanical delay line with a free space beam path (*General Photonics MDL-002*) used as a reference phase sampling technique to characterize the tunability of the DFBs. After the initial characterization, the system operates without the delay stage.

The gain section of DFB2 is pumped with a current in the range of 20–100 mA while DFB1 is operated at a constant current of 30 mA and no current is applied to the heaters. At different gain currents of DFB2, the photocurrent of the receiver

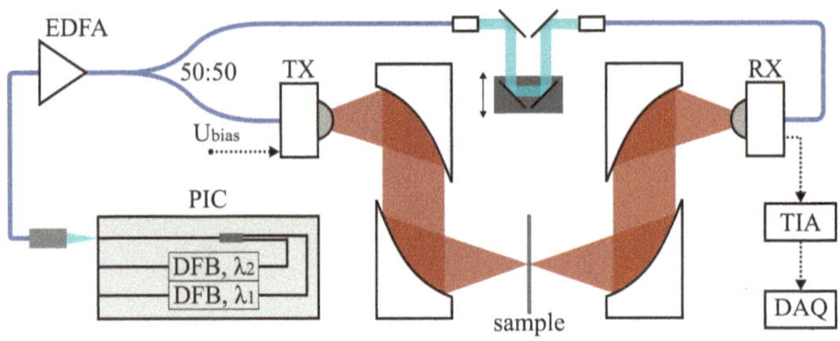

Fig. 14.2 Schematic of the homodyne photonic CW-THz system driven by a PIC

antenna is measured over the position of the delay stage (s. Fig. 14.3a) and the beat frequency is extracted for each gain current of DFB2. The beat frequency is tuned by 400 GHz while the repetitive waviness in Fig. 14.3b could indicate mode transitions within the laser. The same measurements are repeated for heater currents of 0 mA to 30 mA on DFB2 (s. Fig. 14.3c). The available 25 GHz tuning range of the heater agrees with the 0.2 nm shift measured by the optical spectra analysis. In particular the range from 20 mA to 30 mA shows a linear dependence of the beat frequency on the heater current, which makes this range ideal for phase sampling.

After the investigation of the available bandwidth, the characterization of fine tunability of the beat frequency is of interest for the phase sampling with a controlled frequency sweep. For this the current of the gain and heater segments are increased with an increment of 50 µA, which is the minimal step size available by the current source (*Fraunhofer HHI PIConnect*). The photocurrent is measured over delay stage position and gain current as well as delay stage position and heater current. It can be recognized, that phase sampling by tuning the heater current is achievable visualized by the tilted phase picture in Fig. 14.4b. Instead of sampling along the

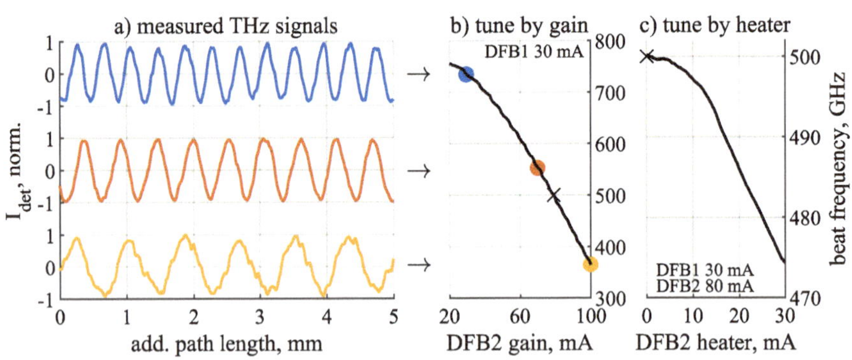

Fig. 14.3 **a** Normalized THz signal versus delay stage position for different beat frequencies: blue 720 GHz, red 560 GHz, yellow 360 GHz. **b** beat frequency versus DFB2 gain current. **c** beat frequency versus heater current of DFB2

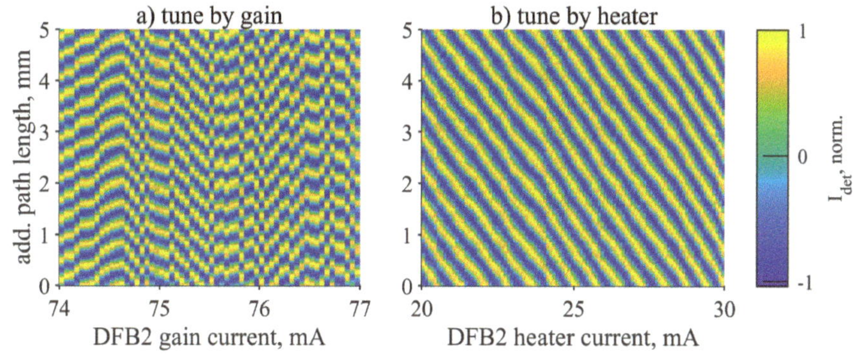

Fig. 14.4 Normalized THz signal versus delay stage position and **a** DFB2 gain current or **b** DFB2 heater current

vertical direction (e.g. in a system with a delay stage) to reconstruct the amplitude and phase of the THz signal, a scan along the horizontal direction can be performed utilizing the heater element. This would not be able with the gain current alone, without the heaters. Figure 14.4a shows a complicated phase picture, where no signal information could be reconstructed in the horizontal direction. This could result from undersampling caused by the current increments of 50 μA.

This system configuration is demonstrated by measurement of material thickness in transmission. ABS samples with different thicknesses are placed in the system shown in Fig. 14.2. The delay stage is replaced by a 1 m optical fiber to ensure an adequate path length difference between both arms. DFB1 is pumped with 30 mA and DFB2 with 86.5 mA while DFB2's heater element is supplied with 26–28 mA resulting in a beat frequency of 451–453 GHz. The sample thickness is calculated by the phase change of the THz signal resulting from a longer optical path induced by the refractive index of the sample. Figure 14.5 shows the measured THz signal for no sample as well as for 300 and 350 μm thick 3D printed ABS samples after sine-fitting and normalization. The measured thicknesses of the samples are calculated from the signal shift $\Delta\theta = 80°$ for the 300 μm sample and 93° for the 350 μm sample to 300.8 μm and 349.7 μm, respectively, via

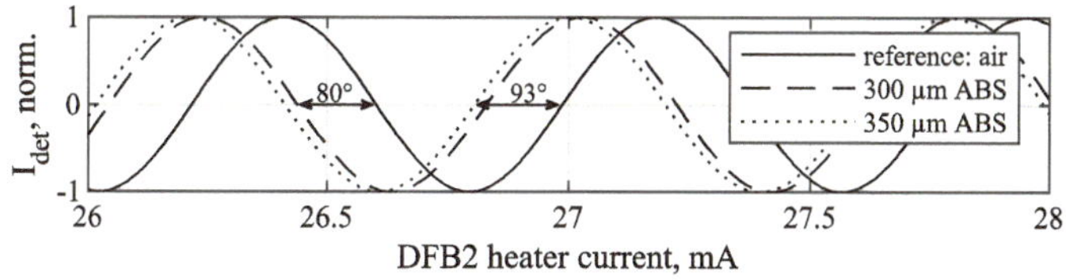

Fig. 14.5 THz signal versus DFB2 heater current measured for different ABS samples

$$d = \frac{\Delta\theta}{2\pi} \cdot \frac{c}{f_{THz}} \cdot \frac{1}{n_{sample} - 1} \tag{14.2}$$

where $f_{THz} = 452$ GHz is the average beat frequency and $n_{sample} = 1.49$ [5].

14.3 Conclusion

In this study, we have demonstrated the potential of using two superimposed distributed feedback (DFB) lasers on a custom-designed PIC as a beat signal source for CW-THz systems. Our experimental results show that the DFB lasers can be tuned over a significant range via gain current, providing a continuous difference frequency range from 200 GHz to 1.2 THz. The homodyne CW-THz system setup successfully demonstrated the phase sampling capability through controlled frequency sweeps of the DFB lasers enabled by the heater elements. Furthermore, accurate measurements of material thickness in transmission was demonstrated using this sampling technique on ABS samples of varying thicknesses.

In conclusion, the integration of DFB lasers on a PIC as the photonic source presents a promising approach towards integrated photonic CW-THz systems. Future work will focus on testing of various configurations of the multipurpose PIC design. Also, the integration of further system components will be addressed in collaboration with Fraunhofer HHI.

Acknowledgements This work was funded by the Deutsche Forschungsgemeinschaft (DFG, German Research Foundation), Project ID 287022738, TRR 196, Project C07.

References

1. Federici, J.F., et al.: THz imaging and sensing for security applications—explosives, weapons and drugs. Semicond. Sci. Technol. **20**, S266–S280 (2005)
2. Jepsen, P.U., Cooke, D.G., Koch, M.: Terahertz spectroscopy and imaging— modern techniques and applications. Laser Photon. Rev. **5**, 124–166 (2011)
3. Mouret, G., et al.: THz media characterization by means of coherent homodyne detection, results and potential applications. Appl. Phys. B **89**, 395–399 (2007)
4. Göbel, T., et al.: Continuous-wave terahertz system with electro-optical terahertz phase control. Electron. Lett. **44**, 863–864 (2008)
5. Squires, A., Lewis, R.: Feasibility and characterization of common and exotic filaments for use in 3D printed terahertz devices. J. Infrared Millim. Terahertz Waves **39**, 1–22 (2018)

Chapter 15
Development, Manufacturing and Characterization of an Integrated Optical Grating for Atomic Interferometers

Sascha de Wall, Alexander Kassner, Folke Dencker, Marc Christopher Wurz, Julian Lemburg, Hendrik Heine, and Ernst Maria Rasel

Abstract High-precision matter wave interferometers offer advantages over classical interferometers in terms of measurement precision, reproducibility and obsolete calibration effort and maintenance. Due to the still stationary construction and the high complexity, the use for commercial purposes is difficult to realize. This requires further miniaturization and integration of the overall system. A possible approach to further reduction of the mass and volume can be the use of a grating MOT (magneto-optical trap). For the previous MOT, four laser beams were needed in the experiments, which had to be adjusted and controlled accordingly. By integrating a micro-engineered grating on the atom chip surface, the number of required laser beams is reduced to one, which illuminates the atom chip surface orthogonally. The focus of this publication is the integration technique as well as the production and characterization of the reflection gratings to achieve the required optical performance. This idea represents an extension of atom chip technology and is the next step towards miniaturization of the overall system.

15.1 Introduction

At the Institute of Micro Production Technology (IMPT) of Leibniz University Hannover, in cooperation with the Institute of Quantum Optics (IQ), atom chips are developed and manufactured which are part of a magneto-optical trap (MOT) for

S. de Wall (✉) · A. Kassner · F. Dencker · M. C. Wurz
Institute of Micro Production Technology, Leibniz University Hannover, Garbsen, Hannover, Germany
e-mail: dewall@impt.uni-hannover.de

J. Lemburg · H. Heine · E. M. Rasel
Institute of Quantum Optics, Leibniz University Hannover, Hannover, Germany

P. Smowton and T. Krauss (eds.), *Proceedings of the 26th European Conference on Integrated Optics*, Springer Proceedings in Physics 440,
https://doi.org/10.1007/978-3-032-29673-3_15

compact matter wave interferometers to measure the absolute gravitational acceleration with high accuracy. In the past, initial investigations have already been successfully carried out in the laboratory and under real conditions. For this purpose, a Bose-Einstein condensate (BEC) was generated in the QUANTUS project [1]. The follow-up project QUANTUS-2 led to a miniaturization of the setup and to testing in drop tower experiments [2]. With the knowledge gained, it was possible to generate a BEC in space within the framework of the MAIUS-1 mission to detect and characterize it by means of atomic interference [3]. In order to make atom chip technology usable for commercial applications, miniaturization and a reduction in complexity must be driven forward. The investigation of novel optical concepts, which reduce the complexity and thus the mass of the lasersystem is therefore necessary. Currently, simple optical gratings are purchased from external suppliers and adhesively bonded to the atom chips. This process leads to inaccuracies in the assembly and also to a high outgassing rate, which limits the vacuum quality and requires corresponding vacuum pumps. In addition, aging and radiation resistance requirements must be considered, especially under space conditions. Furthermore, integration of additional wires close to the chip surface will allow to implement magnetic traps with high trap frequencies to allow for evaporative cooling towards a BEC.

A new approach is to integrate the optical grating into the atom chip as part of the manufacturing process in order to avoid the use of adhesives and ultimately achieve a reduction in the complexity of the overall technology. These gratings have geometric dimensions in the sub-micrometer range and are micro manufactured at the IMPT. Due to the resolution limit of optical photolithography, the structuring is carried out by means of laser beam lithography and the subsequent realization of the gratings by an ion beam etching process.

15.2 Design and Concept

The entire atom chip structure is realized by a stacked multichip-setup, which is accommodated on a ceramic carrier system. To generate strong magnetic fields, mesoscopic wires are used, which are integrated into the carrier. The final magnetic field configurations are generated by the stacked multichip-setup [4]. Therefore, the grating should be integrated close to the surface in order to cool the atoms close to the strongest magnetic field. To visualize the principle setup, Fig. 15.1 shows a schematic overall design.

The focus of this publication is therefore on the production of the science chip with an integrated grating and the associated optical characterization. For a diffraction grating with a corresponding optical quality, a highly polished surface is still required, as is usually found with silicon-based wafers. Due to the embedding of electrical conductor tracks consisting of copper and gold, a chemical-mechanical polishing is necessary for the integration of the grating. With the electrical insulation layers, bonding agent and Si as the base material, this results in a range of materials with different hardnesses on the surface. This leads to uneven removal during polishing, so

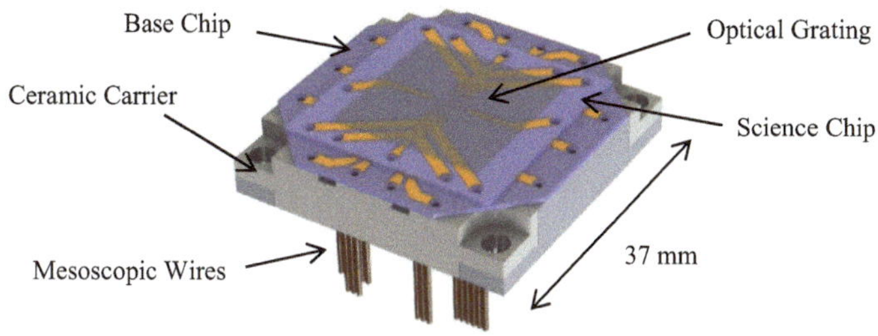

Fig. 15.1 Principle layout of the atom chip system [5]

that the surface quality of highly polished Si is not achieved and further lithographic processes to produce an optical grating in the sub-µm range are not possible.

The idea is therefore to utilize both surfaces of a silicon on insulator (SOI) wafer. An optical grating is integrated on the top surface and deep trenches are etched on the back using DRIE. The buried oxide (BOX) layer in the SOI-wafer serves as an etch stop with a selectivity between SiO_2 and Si of 1:50 so that the conductor tracks lie in one plane. These are coated anisotropic using PVD without coating the sidewalls, thus avoiding the creation of U-shaped conductor tracks. The deposited material not in the trenches is then removed using CMP so that only the shaped conductive tracks deep in the substrate remains. In Fig. 15.2 is a schematically cross-section of the desired System shown.

The integration of the optical grating on the front side takes place by means of sub-µm structures using laser beam lithography. This requires precise adjustment to the back side. After structuring the resist, the grating structure is transferred to the Si by ion beam etching. In later operation, a laser diameter of 20 mm with a wavelength of 780 nm will be used, so the total area of the grating must also be in this order of magnitude. In order to substitute the beam configuration currently in use with four laser beams, the grating design should emerge three partial beams from 1st order diffractions, see Fig. 15.3. With the incident primary beam, the similar configuration

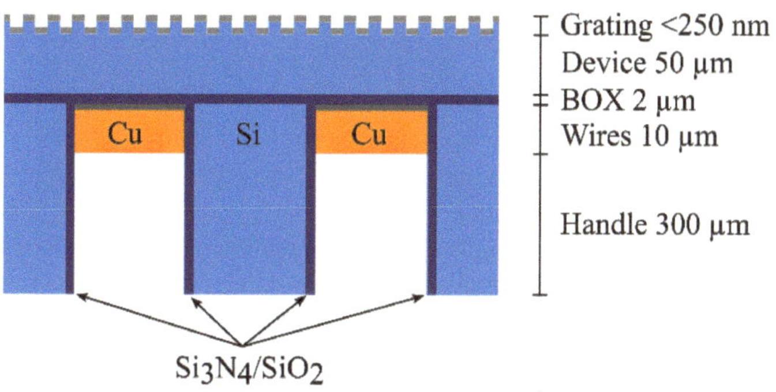

Fig. 15.2 Schematically design of the integrated grating on the atom/science chip

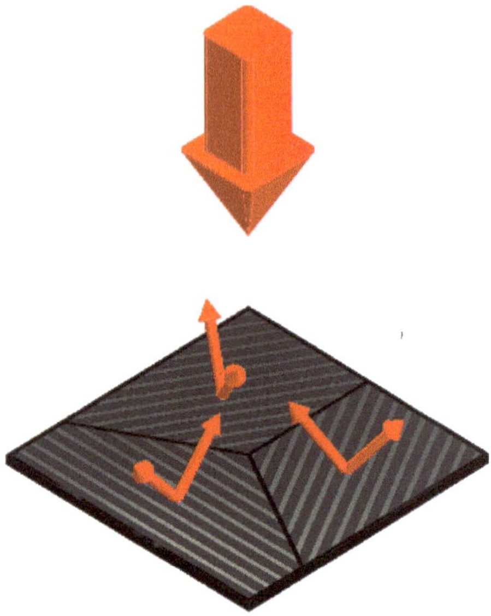

Fig. 15.3 Schematic view of the diffracted beams of the tri-grating

of four beams thus remains [6–8]. Therefore, three 1d Gratings are designed, which are rotated in 120° degrees to each other. The angle α for the 1st diffraction order is chosen to be 45° and the following equation is used to calculate the grating constant g at 1100 nm.

$$g = \lambda / \sin(\alpha) \tag{1}$$

In order to reduce or even eliminate zero-order diffraction as far as possible, a duty cycle of 50:50 in combination with a grating height of 195 nm is aimed for which is λ/4. A final 50 nm coating through PVD with aluminum is applied for the desired reflectivity of approx. 90% at the used laser wavelength.

15.3 Grating Development and Characterization

As already mentioned, laser lithography is used due to the dimensions to structure the grating in a photoresist and afterwards transferred through ion beam etching into the silicon substrate. A final aluminum coating ensures reflective properties. During the production of the first grating, however, it became apparent that the optical properties deviated from the assumptions made in Chap. 2. Although the performance of the 1st diffraction order was as expected in terms of angle and intensities in both directions each with >30% power of the incoming beam, but the 0th order diffraction was significantly stronger than initially assumed. Figure 15.4 shows a cross-section image

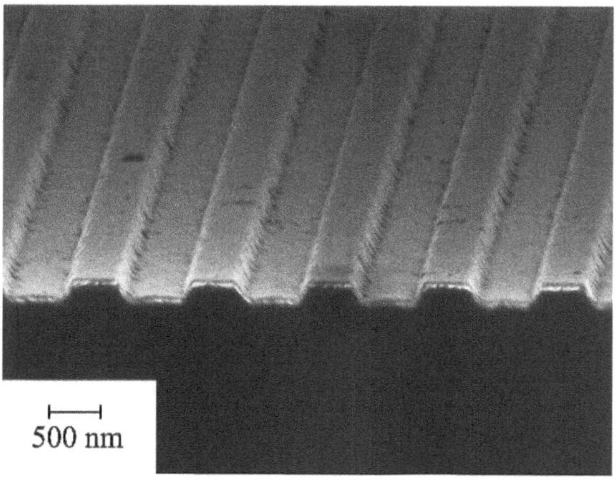

Fig. 15.4 Cross-section (SEM) view of the manufactured grating with a height of 195 nm

for checking the dimensions using a scanning electron microscope and Table 15.1 shows the associated optical parameters.

The design of the grating height by simply assuming only a λ/4 shift appears to be insufficient. Effects such as the aluminum coating, natively aluminum oxide layer, shape fidelity, edge geometry, edge roughness and edge angle due to the manufacturing process have a significant influence on the grating performance [9]. To identify optimal parameters, 1d gratings are first manufactured with an area of 5 mm × 5 mm and various depths, which are large enough to measure the optical performance. Then three 1d gratings are combined to a tri-grating configuration with a total area of 20 mm × 20 mm. For the variation of the grating height, 215 and 235 nm are set to see whether the optical performance changes to the 0th diffraction order. For this purpose, only the IBE process is adjusted in time to achieve the desired height. The test gratings are all etched in parallel with one height on one wafer. In this context, the parameters of the laser exposure system are also varied in order to identify the optimum. The key parameters are the focus offset and the laser power. The results are exemplary shown for the gratings with a height of 235 nm in Fig. 15.5.

From these results, the optimum exposure parameters were already identified in advance by visual control between a focus offset of −15 and −20% and a laser power between 60 and 70 mW. This coincides with the optimum of the 0th diffraction order in this range. The mathematically calculated mean value for the four best performing gratings with regard to the 0th diffraction order is 7.35% at a grating height of ~215 nm and 5.38% at ~235 nm. In Fig. 15.6 cross-section views of the best gratings with a height of ~215 and ~ 235 nm are shown.

Table 15.1 Optical parameters for the manufactured grating with a height of 195 nm

Input (mW)	+1st order	−1st order	0th order
3.89	1.31 mW (33.7%)	1.29 mW (33.2%)	672 μW (17.3%)

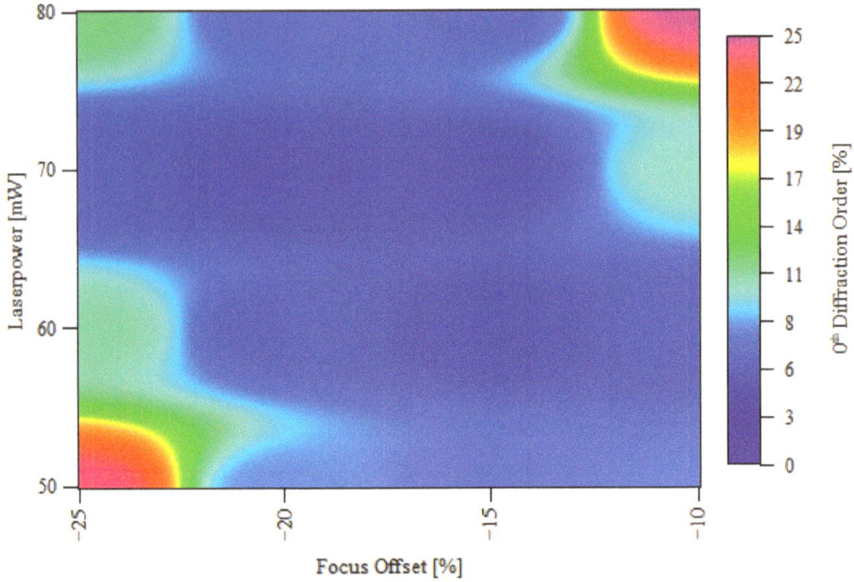

Fig. 15.5 Results of the optical characterization of the test gratings regarding 0th diffraction order

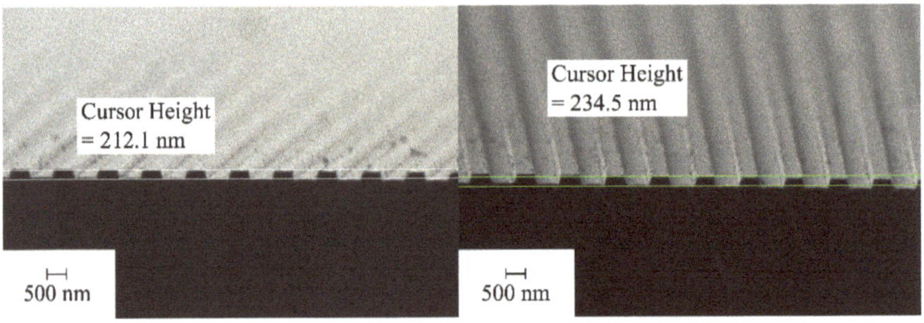

Fig. 15.6 Cross-section (SEM) views of the manufactured gratings with a height of ~215 nm (left) and ~ 235 nm (right)

In the next step, the final grating is manufactured in a tri-grating configuration on the science chip. Figure 15.7 shows a picture of the front and back of the science chip with successfully grating integration and Table 15.2 shows the corresponding characterizations of the individual substructures of the grating.

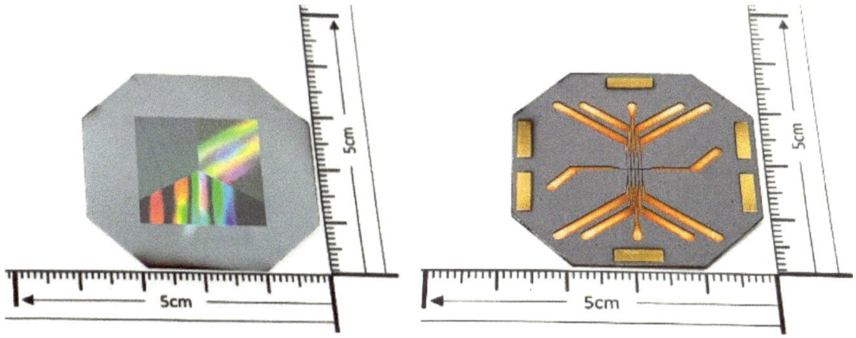

Fig. 15.7 Front (left) and backside (right) of the manufactured science chip with an integrated optical grating

Table 15.2 Results of optical grating in tri-grating formation

	Input (mW)	+1st order	−1st order	0th order
1	4.7	1.44 mW (30.6%)	1.72 mW (36.6%)	157 μW (3.3%)
2	4.83	1.45 mW (30%)	1.53 mW (31.7%)	186 μW (3.8%)
3	4.85	1.59 mW (32.8%)	1.41 mW (29.1%)	224 μW (4.6%)

15.4 Conclusion and Outlook

It was possible to successfully implement the intended design, which leads to miniaturization of the overall system. Furthermore, the grating integration does not require any adhesive bonding agents anymore and the distance between the conductive tracks and the final BEC can be reduced by a factor three with this design, because the substrate thickness of the grating chip is eliminated. This leads to a ten-fold increase in trap frequency. However, the optical performance of the gratings had to be determined iteratively and empirically because there is a discrepancy between the design and the final optical properties especially in the 0th diffraction order. The simplified assumption of a phase shift due to the height difference of the grating unfortunately does not cover all physical influences that occur. In future designs of new gratings, further physical aspects such as sidewall roughness, sidewall angle and material-specific parameters, which can vary depending on the deposition processes, must taken into account.

Acknowledgements This work is supported by the German Aerospace Center with funds from the Federal Ministry of Economic Affairs and Climate Action due to the enactment of the German Bundestag under the funding code 50RK1979 (QCHIP).

References

1. van Zoest, T., Gaaloul, N., Singh, Y., Ahlers, H., Herr, W., Seidel, S.T., Ertmer, W., Rasel, E., Eckart, M., Kajari, E., et al.: Bose–Einstein condensation in microgravity. Science. **328**(5985), 1540–1543 (2010)
2. Rudolph, J., Herr, W., Grzeschik, C., Sternke, T., Grote, A., Popp, M., Becker, D., Muentinga, H., Ahlers, H., Peters, A., Laemmerzahl, C., Sengstock, K., Gaaloul, N., Ertmer, W., Ra-sel, E.M.: A high-flux BEC source for mobile atom interferometers. New J. Phys. **17**(6), 065001 (2015)
3. Becker, D., Lachmann, M.D., Rasel, E.M., et al.: Space-borne Bose–Einstein condensation for precision interferometry. Nature. **562**, 391–395 (2018)
4. Seidel, Stephan Tobias: Eine Quelle für die Interferometrie mit Bose-Einstein-Kondensaten auf Höhenforschungsraketen. Dissertation, Leibniz Universitaet Hannover, Technische Informationsbibliothek (TIB) (2014)
5. Kassner, A., Dencker, F., Künzler, C., Heine, H., Herr, W., Christ, M., Krutznik, M., Rasel, E.M., Wurz, M.C.: Atomchips mit integrierten optischen Gittern zur Erzeugung von Bose-Einstein-Kondensaten. Mikro-SystemTechnik Kongress, Berlin (2019)
6. Barker, D.S., Elgee, P.K., Sitaram, A., Norrgard, E.B., Klimov, N.N., Campbell, G.K., Eckel, S.: grating magneto-optical traps with complicated level structures. New J. Phys. **25**(10), 103046 (2023). https://doi.org/10.1088/1367-2630/ad02ea
7. Franssen, J.G.H., de Raadt, T.C.H., van Ninhuijs, M.A.W., Luiten, O.J.: Compact ultracold electron source based on a grating magneto-optical trap. Phys. Rev. Accel. Beams. **22**(2) (2019). https://doi.org/10.1103/PhysRevAccelBeams.22.023401
8. Nshii, C.C., Vangeleyn, M., Cotter, J.P., Griffin, P.F., Hinds, E.A., Ironside, C.N., et al.: A surface-patterned chip as a strong source of ultracold atoms for quantum technologies. Nat. Nanotechnol. **8**(5), 321–324 (2013). https://doi.org/10.1038/nnano.2013.47
9. Burrow, O., Fasano, R., Brand, W., Wright, M., Li, W., Ludlow, A., Riis, E., Griffin, P., Arnold, A.: Optimal binary gratings for multi-wavelength magneto-optical traps. Opt. Express. **31**, 40871–40880 (2023)

Chapter 16
Accelerated Reliability Testing for Integrated Photonic with Machine Learning Methods

Léa Chaccour, Kristif Prifti, Joep Nieuwdorp, Luc Augustin, Ksenia Abrashitova, Ben Hasenson, and Sylwester Latkowski

Abstract The reliability prediction of Indium Phosphide (InP) Fabry-Pérot (F-P) lasers using machine learning (ML) techniques is investigated. Ensuring the longevity and performance stability of these lasers is critical for their integration into photonic integrated circuits (PICs). Traditional reliability assessment methods require extensive stress testing, which is both time-consuming and costly. To address this, we evaluate various ML algorithms to determine their effectiveness in predicting device degradation based on initial device parameters. The theoretical models are validated using experimental datasets collected from discrete components fabricated on InP-based platform. The results demonstrate that ML models have the potential to reduce the need for prolonged stress testing while maintaining predictive accuracy. The applicability of these ML techniques can be extended beyond the F-P lasers to more integrated photonics devices, offering a promising pathway for cost-effective reliability prediction of volume manufactured photonic integrated circuits.

L. Chaccour (✉) · B. Hasenson · S. Latkowski
PITC, Eindhoven University of Technology (TU/E), Eindhoven, The Netherlands
e-mail: l.chaccour@tue.nl

K. Prifti
VTEC Lasers & Sensors, Eindhoven, The Netherlands

J. Nieuwdorp · L. Augustin · K. Abrashitova
SMART Photonics: High Tech Campus, Eindhoven, The Netherlands

L. Chaccour · L. Augustin · B. Hasenson · S. Latkowski
Photonic Integration Group, Eindhoven University of Technology (TU/E), Eindhoven, The Netherlands

P. Smowton and T. Krauss (eds.), *Proceedings of the 26th European Conference on Integrated Optics*, Springer Proceedings in Physics 440,
https://doi.org/10.1007/978-3-032-29673-3_16

16.1 Introduction

Indium Phosphide (InP) is a mature technology widely utilized in the fabrication of lasers, photodiodes, and waveguides [1]. InP based technologies are among the major platforms for photonic integrated circuits (PICs) [2]. One key advantage of InP is its capability to integrate active and passive components, such as lasers, semiconductor optical amplifiers (SOAs), photodetectors, modulators, waveguides, splitters, combiners, and filters, onto a single monolithic chip. Notably, the InP-based technologies offer efficient and high performing sources, detectors, and modulators in both the C- and O-bands [3], making it an ideal choice for minimizing the cost, size, weight, and power consumption of integrated photonic systems [4]. The reliability of integrated circuits (ICs) is a critical aspect of the microelectronics industry, as it determines a device's ability to consistently perform its intended function over time without failure [5]. With the rapid advancements in photonic IC technologies, there is an increasing demand for comprehensive reliability studies to identify potential failure mechanisms, extend device lifespan, and ensure stable performance under varying conditions. However, traditional aging and stress tests are costly and time-consuming [6]. Machine learning (ML) techniques have the potential to revolutionize PIC reliability studies by enhancing prediction accuracy, reducing testing durations, and enabling early failure detection. This approach can lead to more reliable and efficient PICs, ultimately improving the scalability of integrated photonic systems.

16.2 Machine Learning Model for Predicting Fabry-Pérot Laser Reliability

16.2.1 Experimental Setup and Feature Engineering

In this study, an ML model was used to assess the reliability of Fabry-Pérot (F-P) lasers and reduce the need for extensive testing in the production flow. The F-P lasers, obtained from different locations on two wafers and with varying lengths of the gain sections (L_G) (300, 500, and 750 μm), are measured after dicing. The parameters under consideration include gain section length (L_G), laser location on the wafer (Loc_w) which is the distance from the center of the wafer, and lasing threshold current I_{th}. The threshold current I_{th} reflects the laser efficiency and performance. It is the current at which stimulated emission inside the laser cavity begins to dominate over spontaneous emission, marking the onset of lasing. For a given laser type, higher threshold often indicate increased losses or reduced internal quantum efficiency (η_{int}), which measures the efficiency of converting injected carriers into photons in the active region. Increased I_{th} values can be associated with active region aging or degradation, as these processes increase non-radiative recombination and other loss mechanisms. To predict reliability, we are focusing on the internal efficiency changes of the lasers, the device lifetime is determined based on the failure threshold current. This approach helps to understand when the lasers will fail due to efficiency changes over time.

New features were created and added to the model, to capture key factors affecting the laser's performance, such as cumulative stress (C_s), length interaction (L_i) (Cavity Length L_G * Location on the wafer Loc_w), and cavity length and threshold current interaction, as shown in Eqs. (16.1), (16.2), and (16.3), with I_{th0} representing the threshold current at the 0 h readout milestone M_0.

$$C_S = \text{Temperature stress} * \text{Current stress} * \text{Milestone} \tag{16.1}$$

$$L_i = L_G * Loc_w \tag{16.2}$$

$$\text{Cavity length interaction} = L_G * I_{th0} \tag{16.3}$$

74 devices were initially tested at milestone M_0 using Light–Current-Voltage (LIV) characteristics measured at 25 °C. They were then subjected to thermal stress at 85 °C and current stress under a current density of 15 kA/cm^2 for 200 circuits and 8 kA/cm^2 for 138 circuits. After this stress period, measurements were conducted again at 25 °C to collect data at the 24-h milestone, M_{24}. Notably, the tests at each milestone were carried at room temperature (25 °C), on a separate test bench and not in-situ within the climate chamber. This process was repeated at subsequent milestones, M_{48}, M_{72}, M_{168}, and $M_{520/600}$ h. For the ML model, we chose features that allow the model to predict I_{th} from I_{th0} at a certain milestone and these were validated experimentally up to 600 h.

According to the Telcordia SR-332 (Issue 4) document, the reliability prediction procedure for electronic component is described in detail; however, the I_{th} failure percentage is not explicitly specified, and implemented independently at a company or entity level. At any given milestone, the time to failure is defined as the point where I_{th} exceeds a specified percentage of its initial value (I_{th0}). In the literature this percentage is stated to be 50% of I_{th0} [7]. Consequently, we used I_{th} as the target variable and included C_s, L_i and cavity length interaction as features. By incorporating these parameters, the model can predict I_{th} evolution over time. At any given milestone, if I_{th} exceeds a certain percentage of its initial value I_{th0}, the corresponding time is considered the time to failure. The model establishes a relationship between I_{th} and the independent parameters, thereby facilitating effective reliability predictions. This approach should help to reduce the need for time consuming excessive testing at larger milestones.

16.2.2 Model Selection and Performance Evaluation

To get the best predictive results, we iterated over several models [8], including XGBoost, Linear Regression (LR), Artificial Neural Network (ANN), Gradient

Boosting (GB), Support Vector Regression (SVR), Ridge Regression (Ridge), and Random Forest (RF). The models were evaluated using R^2 (R-squared) and RMSE (Root Mean Squared Error), with R^2 demonstrating how well the model accounts for data variability, and RMSE indicating the average size of the prediction errors. A high R^2 value (close to 1) and low values of RMSE (close to 0) indicate a well-performing model with accurate predictions. In our case, tree-based models (XGBoost, GB and RF) performed well overall. These models are machine learning algorithms that use a hierarchical structure of decision trees to make predictions. They work by recursively splitting the data into subsets based on specific variables and making predictions using the majority class (for classification) or the average value (for regression) within each subset. Tree-based models often outperform deep learning models on tabular data, as is the case in our study [9]. Fine-tuning of the hyperparameters is essential to achieve optimal performance in such models. The hyperparameters that were tuned include the number of estimators (number of trees), learning rate (step size for updates), maximum depth (tree complexity), subsample size (fraction of data used per tree), and minimum samples required for a split (node-splitting threshold), to optimize its performance. These parameters were carefully adjusted to optimize the model's performance and improve prediction accuracy [10]. Figure 16.1 shows high R^2 and low RMSE for both test and training data, further confirming the good performance of the Gradient Boost model (GB). Negative values of R^2 in Fig. 16.1, denotes that the machine-learning model fits the data poorly [11]. With our data from just two wafers, there's a risk of overfitting. Cross-validation can help, but it doesn't completely solve the problem. This issue can be mitigated in the future once more data is introduced. To boost the model's robustness to unseen data, we used K-fold cross-validation [12] (However, the number of data and it's variety isn't enough to ensure generalizability). This data validation technique partitions the dataset into K subsets (folds), where each fold is used for testing once, while the remaining K-1 folds are used to train the model. The final performance metrics are then averaged across all folds. We compared the R-squared and RMSE for both 5-folds and 10-folds, we chose 5 and 10 folds because they provide a good balance between computational efficiency and model performance estimation. No significant difference was found between the 5 and 10-folds, indicating comparable stability and performance [12]. To evaluate the model's accuracy, a portion of the dataset was set aside as test data, 20% of the data was used for model testing. Figure 16.2 presents the predicted I_{th} values at different milestones for two F-P lasers from two different wafers, compared to the actual I_{th} values. The calculated relative error is also displayed in the same figure. The relative error is defined as the absolute difference between the predicted and actual values, divided by the actual value, and expressed as a percentage. These results show that the model predicts I_{th} up to a milestone of 600 h. The error bar represents the variability in the model predictions, with different errors arising from factors such as model approximation and feature relationships. Our machine learning model provides useful insights, but comparing it directly to real-world devices is challenging due to limited on-die data. Additionally, differences in industrial packaging can affect reported I_{th} values, making real-world comparisons more complex.

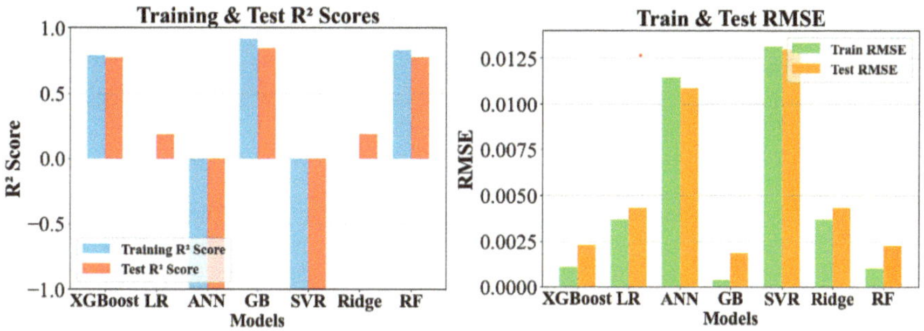

Fig. 16.1 R^2 score and RMSE comparison for different models: XGBOOST, LR (Linear Regression), ANN (Artificial Neural Network), GB (Gradient Boosting), SVR (Support Vector Regression), Ridge Regression (Ridge) and RF (Random Forest)

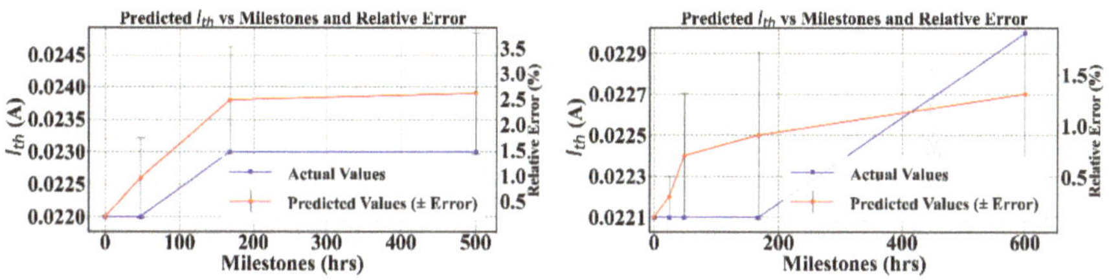

Fig. 16.2 Figure showing the predicted I_{th} values using the Gradient Boosting model for two types of integrated F-P lasers, with the relative error displayed on the right y-axis and the I_{th} of actual and predicted values on the left y-axis

16.3 Conclusion

We used a machine learning model and trained it to predict the reliability of InP F-P lasers by estimating I_{th} at future milestones, reducing the need for extensive Light–Current–Voltage (LIV) testing at high temperatures and current stresses. Among the models evaluated, the Gradient Boosting model demonstrated the best performance after hyperparameter optimization, confirming its effectiveness for reliability prediction. This study suggests that machine learning can contribute to streamlining the reliability testing process by providing accurate predictions. Future work, will focus on testing the model under various temperature and stress conditions to assess its robustness across diverse operating scenarios. Additionally, extending this approach to other integrated components could further optimize reliability testing and improve predictive accuracy, positioning machine learning as a transformative tool in the development of integrated photonic circuits and components.

Acknowledgements This project has received funding from the PhotonDelta PIC-Test programme.

We would like to acknowledge Dzmitry Pustakhod for his valuable contribution to this work. His insights and assistance were greatly appreciated.

References

1. Arafin, S., Coldren, L.: Advanced InP photonic integrated circuits for communication and sensing. IEEE J. Sel. Top. Quantum Electron. **24** (2018)
2. Smit, M., et al.: An introduction to InP-based generic integration technology. Semicond. Sci. Technol. **29** (2014)
3. MPW Run Schedule. https://www.jeppix.eu/workflow/mpw-run-schedule/
4. Klamkin, J., Zhao, H., Bowen, S., Liu, Y., Isaac, B., Pinna, S., Sang, F., Coldren, L.A.: Indium phosphide photonic integrated circuits: technology and applications. In: Presented at the Proceedings of the 2018 IEEE Boston Conference on IC and System-on-Chip (BCICTS), Boston, USA (2018)
5. Naresky, J.: Reliability definitions. IEEE Trans. Reliab. R-19 (1970)
6. Ikegami, T., Fukuda, M., Suzuki, M., Considerations on the degradation of DFB lasers. In: Semiconductor Device Reliability. Springer, Dordrecht (1990)
7. Yue, S., Zhiyong, L., Jiaming, B., Shen, N., Zibo, W., Li, Q., Yongyi, C., Lei, L., Yuxin, L., Peng, J., Xiaonan, S., Lijun, W.: Processes of the reliability and degradation mechanism of high-power semiconductor lasers. Crystals **12** (2022)
8. Hearty, J.: Advanced Machine Learning with Python (2016)
9. Grinsztajn, L., Oyallon, E., Varoquaux, G.: https://arxiv.org/abs/2207.08815
10. Li, H.: Machine Learning Methods (2024)
11. Dutta, A.: A negative value of R-square in the data science world—is it a myth? Medium. https://medium.com/data-science-insights-and-predictions/a-negative-value-of-r-square-in-the-data-science-world-is-it-a-myth-fa6488e69efc
12. Kohavi, R.: A study of cross-validation and bootstrap for accuracy estimation and model selection. In: IJCAI'95: Proceedings of the 14th International Joint Conference on Artificial Intelligence, Montreal, Canada (1995)

Chapter 17
Low Loss Single Arm Bimodal/Trimodal Gold-Si_3N_4 Plasmo-Photonic Sensor

Panagiotis Zdoupas and Konstantinos Vyrsokinos

Abstract This work presents the design and optimization of a high-performance plasmo-photonic interferometer on the Si_3N_4 platform, employing both aluminum (Al) and gold (Au) plasmonic waveguides. The Al-based sensor achieves a bimodal interference configuration, delivering a sensitivity of 10,340 nm/RIU and an Extinction Ratio (ER) of up to 39 dB, rendering it ideal for applications requiring high sensitivity and easy resonance detection. In contrast, the Au-based sensor exhibits trimodal interference under the same configuration, resulting in increased ER at the cost of sensitivity. Through systematic optimization of the plasmonic film thickness, BOX thickness, and photonic taper width, one of the two bottom modes is effectively suppressed, enabling bimodal operation with a sensitivity of 10,500 nm/RIU and an ER of 24 dB. These results demonstrate the trade-offs between sensitivity and ER in plasmonic sensing and establish a design framework for next-generation sensors targeting biosensing, environmental monitoring, and chemical analysis applications.

17.1 Introduction

Plasmo-photonic sensing has emerged as a powerful technology for detecting minute changes in refractive index of gases and liquids, enabling applications in biosensing, environmental monitoring, and chemical analysis [1]. By leveraging the unique properties of surface plasmon polaritons (SPPs), that provide extreme light confinement and enhanced light-matter interactions, these sensors achieve ultra-high sensitivity while maintaining a compact footprint [2]. Plasmonic waveguides, are typically fabricated from noble metals like gold (Au) or silver (Ag), but there is a strong research trend to explore CMOS-compatible materials such as aluminum (Al) or copper (Cu) targeting to address the cost issue for mass-producible sensors [3]. Al in particular

P. Zdoupas (✉) · K. Vyrsokinos
School of Physics, Aristotle University of Thessaloniki, Thessaloniki, Greece
e-mail: pzdoupas@auth.gr

Center for Interdisciplinary Research and Innovaton, Aristotle University of Thessaloniki, Thessaloniki, Greece

P. Smowton and T. Krauss (eds.), *Proceedings of the 26th European Conference on Integrated Optics*, Springer Proceedings in Physics 440,
https://doi.org/10.1007/978-3-032-29673-3_17

has the additional attractive feature of being oxidization resistant, due to an alumina layer that is formed at the plane's surface [4]. However, while Al-based plasmonic sensors exhibit high sensitivity and strong extinction ratios (ER) [5], they suffer from significant propagation losses compared to Au, limiting the maximum sensing length before the output power drops below the detection threshold of the read-out system.

This work presents through numerical simulations, an optimized single-arm plasmo-photonic interferometer on the Si_3N_4 platform, employing Au plasmonic waveguides. This design targets reduced propagation losses, while enhancing metal surface functionalization compared to Al. However, transitioning from Al to Au is not straightforward, as two-dimensional (2D) simulations revealed that two modes propagate with nearly equal power at the bottom metal–dielectric interface, resulting in trimodal operation. Under these conditions, the interference pattern maintains high ER but exhibits reduced sensitivity. By optimizing the geometric parameters of the photonic-to-plasmonic interface and adjusting the Au thickness, one of the two bottom modes can be sufficiently suppressed, enabling nearly pure bimodal interference. The final optimized structure achieves a sensitivity of 10,500 nm/RIU and an extinction ratio of up to 24 dB. By addressing these key challenges in propagation losses and modal interference, this work proposes a highly sensitive, scalable, and functionalized sensing platforms, paving the way for breakthroughs in biosensing, environmental monitoring, and beyond.

17.2 Results

Figure 17.1a shows a bird's-eye view of the proposed structure. The device features a Si_3N_4 waveguide with a thickness of 0.8 μm, tapered from 1 to 7 μm. The top and bottom claddings are SiO_2, while the 80 nm thick metal film is either from Au or Al. The light is injected from the photonic to the plasmonic waveguide and vice versa via butt coupling. In this configuration the plasmonic waveguide supports one top mode serving as the sensing actuator, while the bottom one acts as the reference. However, from the overall integral calculations presented in Fig. 17.1b it is revealed that this structure supports also a third-order bottom SPP mode. For Al, the power of this third-order mode remains relatively low compared to the first-order bottom and top SPP modes being less than 4% in terms of coupled power. Combined with the high losses of the third order-bottom SPP mode from Fig. 17.1c, it is clear that the output spectrum will result from the bimodal interference of these two modes. Figure 17.1d–f show the |Ey| distribution of the 1st order top, 1st order bottom and 3rd order bottom SPP mode, respectively. The comparison of the mode field intensity between the Fig. 17.1e, f confirms that the 1st order bottom mode is significant more intense than the 3rd order bottom one. On top of that, it is evident that the 3rd order bottom mode is much less confined within the metal-BOX interface, leading to the conclusion that it's a free-space mode.

However, when the SPP waveguide's metal is replaced by Au of the same thickness, according to Fig. 17.1b the 3rd order bottom mode gains significant more

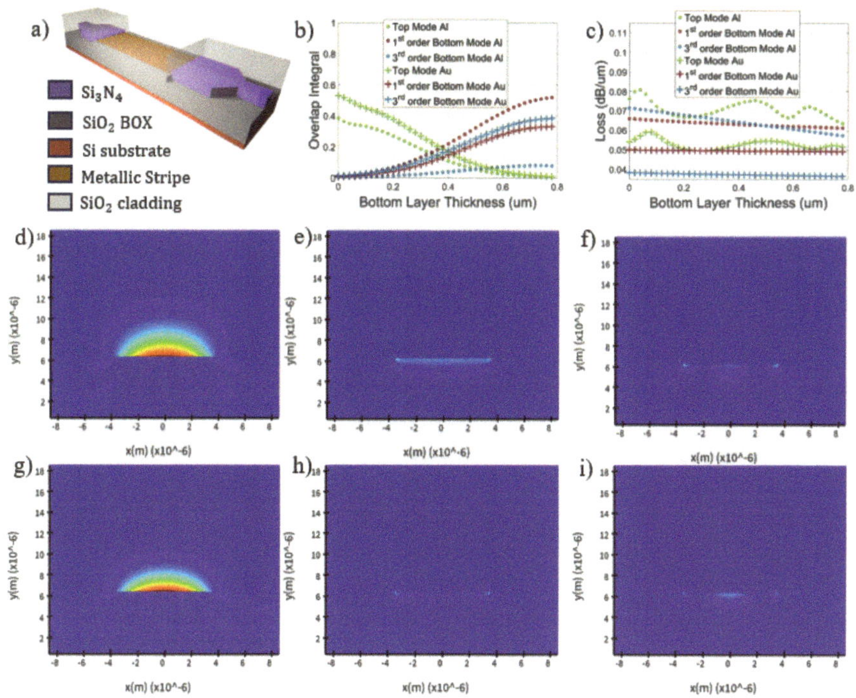

Fig. 17.1 **a** Bird view of the proposed plasmo-photonic sensor, **b** Overlap integral of the three supported SPP modes for Al & Au, **c** Propagation Losses of the three supported SPP modes for Au & Al, |Ey| distribution of the **d** 1st order top SPP mode, **e** 1st order bottom SPP mode, **f** 3rd order bottom SPP mode for Al plasmonic metal, **g** Ey| distribution of the **h** 1st order top SPP mode, **d** 1st order bottom SPP mode, **i** 3rd order bottom SPP mode for Au plasmonic metal

power from the higher coupling coefficients. This is confirmed also by the results of Fig. 17.1h, i displaying the |Ey| distribution of the 1st order bottom and 3rd order bottom SPP modes, while the 1st order top mode maintains the same field profile evident by Fig. 17.1g.

Figure 17.2a presents the resulting output spectra under these conditions generated by 3D-FDTD simulations with a resolution of 0.5 nm across the entire wavelength span of 1300–1600 nm. The sensitivity is calculated by changing the surrounding medium on top of the metal plasmonic surface with two different water solutions assuming refractive indices equal to 1312 and 1312. Then from the resonance wavelength shift the sensitivity is calculated from the simple equation:

$$S = \frac{\Delta\lambda_{res,peak}}{\Delta n} \tag{17.1}$$

Under this trimodal interference, the achievable extinction ratio (ER) at the output of the interferometer is reduced, as illustrated in Fig. 17.2a, where the median ER is

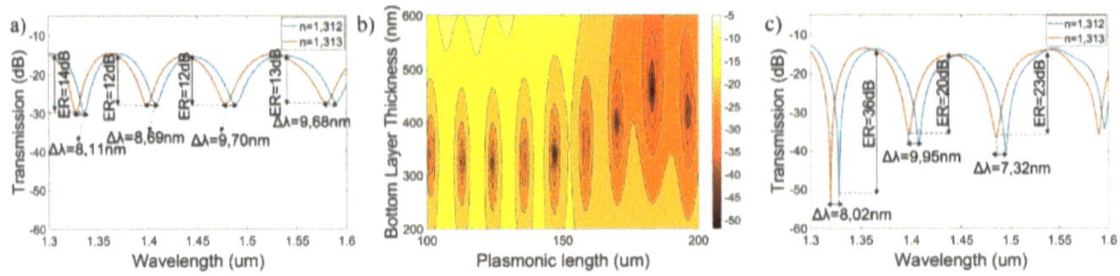

Fig. 17.2 **a** Un-optimized Al-based trimodal interferometer spectral response, **b** Output power of interferometer at 1550 nm for variable bottom layer thickness and plasmonic length, **c** Spectral response of the Al-based optimized trimodal configuration

approximately 13 dB. The maximum sensitivity ranges between 9700 and 9680 nm/RIU for the two resonance peaks at 1480 nm and 1540 nm, respectively.

In order to identify the maximum ER for this Au based sensor, the output power of the interferometer was scanned by varying the BOX thickness as well as the plasmonic length, with the corresponding results displayed in Fig. 17.2b. The minimum power was recorded for 147 um long waveguide and a BOX thickness of 0.34 um. Figure 17.2c displays the calculated spectral response for this layout for the two top liquid claddings. The sensitivity of the sensor was measured equal to 8020 nm/RIU at the peak of 1320 nm and an ER of 36 dB. This sensitivity value is only 76% of its bimodal counterpart in terms of sensitivity for the best obtainable ER for this trimodal sensor. For the equivalent resonance peak at 1500 nm the sensitivity is measured 7320 nm/RIU and the ER is 23 dB. Therefore, the Au trimodal sensor exhibits inferior performance to the Al based one.

Achieving a high ER above 20 dB in combination with sensitivities exceeding 10,000 nm/RIU requires suppressing one of the two propagating bottom modes. To address this, the plasmonic film's thickness was scanned across a range of 30–200 nm, identifying 50 nm as the minimum thickness capable of supporting the SPP bottom modes. Figure 17.3a illustrates these results, showing that for the minimum 50 nm Al thickness, the first-order bottom mode is effectively suppressed for BOX thicknesses above 320 nm, with a coupling coefficient of approximately 5%. Furthermore, as depicted in Fig. 17.3b, reducing the photonic taper width from 7 to 3 μm decreases the coupling coefficient of the first-order bottom mode to almost 0%, resulting in a pure bimodal configuration where only the top and third-order bottom modes interfere at the output. Figure 17.3c presents the 3D FDTD simulations for the two liquids on top of the plasmonic surface. The resulting spectra reveal that for a 340 nm BOX thickness, the sensitivity at the peak near 1530 nm reaches 10,500 nm/RIU, with an ER of 24 dB. For the resonance peak at 1355 nm, the ER increases to 42 dB, but the sensitivity drops to 8950 nm/RIU. These findings highlight the trade-offs between sensitivity and ER, providing valuable insights for optimizing plasmonic sensor performance.

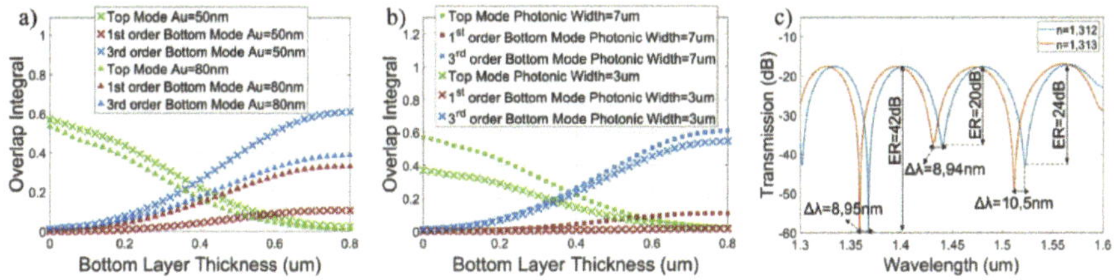

Fig. 17.3 **a** Overlap integral for Au film thickness equal to 50 and 80 nm, **b** Overlap integral for photonic waveguide width equal to 7 and 3 um, **c** Spectral response of the optimized bimodal Al sensor

17.3 Conclusions

In conclusion, this work demonstrates the design and optimization of a high-performance plasmo-photonic interferometer on the Si_3N_4 platform, employing both aluminum (Al) and gold (Au) plasmonic waveguides. The Al-based sensor providing CMOS compatibility achieves a bimodal interference configuration, yielding a sensitivity of 10,340 nm/RIU and an extinction ratio (ER) of up to 39 dB, rendering it highly suitable for applications requiring high sensitivity and strong signal contrast. However, the Au-based sensor targeting lower propagation losses and improved surface functionalization, exhibits trimodal interference due to the enhanced coupling of the third-order bottom mode, resulting in reduced sensitivity and ER compared to the Al counterpart. Through systematic optimization of the plasmonic film thickness, BOX thickness, and photonic taper width, the third-order mode is effectively suppressed, restoring a near-bimodal operation with a sensitivity of 10,500 nm/RIU and an ER of 24 dB. These results highlight the critical trade-offs between sensitivity and ER in plasmonic sensing paving the way for advanced applications in biosensing, environmental monitoring, and chemical analysis.

References

1. Divya, J., et al.: Surface plasmon based plasmonic sensors: a review on their past, present and future. Elsevier Biosens. Bioelectron. X **11** (2022)
2. Gao, Y., Xin, Z., Gan, Q., Cheng, X., Bartoli, F.J.: Plasmonic interferometers for label-free multiplexed sensing. Opt. Express **21**, 5859–5871 (2013). https://doi.org/10.1364/OE.21.005859
3. Weeber, J. et al.: Characterization of CMOS metal based dielectric loaded surface plasmon waveguides at telecom wavelengths. Opt. Exp. **25**, 394–408 (2017)
4. Langhammer, I., et al.: Localized surface plasmon resonances in aluminum nanodisks. Nano. Lett. **8**, 1461– 1471 (2008). https://doi.org/10.1021/nl080453i
5. Fotiadis, K., et al.: High-sensitivity bimodal plasmo-photonic refractive index sensor. ACS Photonics **10**(8), 2580–2588 (2023)

Chapter 18
Simulation and Design PDK for LNOI PIC Platform

Andrzej Polatynski, Florian Dubois, Gilles Feugnet, Waldemar Schmidt, Homa Zarebidaki, Jacopo Leo, Hamed Sattari, Alexandre Shen, Cosimo Calò, and André Richter

Abstract The rapid advancement of photonic integrated circuit (PIC) technologies has driven the development of novel material platforms with enhanced performance. Thin film lithium niobate on insulator (LNOI or TFLN) has emerged as a highly promising PIC platform due to its exceptional electro-optic properties, strong nonlinearities, and wide transparency window. However, the commercial adoption of LNOI remains limited by the absence of standardized process design kits (PDKs), which are essential for scalable design and fabrication. To address this challenge, we present a simulation PDK building block library for LNOI PICs, developed within the H2020-ELENA project and tailored for an open PIC foundry supported by Swiss Center for Electronics and Microtechnology (CSEM). This PDK enables realistic circuit-level simulations, accounting for fabrication constraints and technology-specific design rules. To validate the functionality of the PDK, we demonstrate three end-user prototypes: a linearized electro-optic modulator for RF signal transmission, a high-speed, widely tunable laser for telecom applications, and a frequency shifter for quantum sensing. These prototypes integrate various LNOI building blocks, especially phase shifters, PPLN waveguides, and high-speed optical modulators. Our work establishes a robust foundation for the scalable development of LNOI-based PICs, bridging the

A. Polatynski (✉) · A. Richter
VPIphotonics GmbH, Berlin, Germany
e-mail: andrzej.polatynski@vpiphotonics.com

A. Polatynski
Warsaw University of Technology, Warsaw, Poland

F. Dubois · H. Zarebidaki · J. Leo · H. Sattari
Swiss Center for Electronics and Microtechnology (CSEM), Neuchâtel, Switzerland

G. Feugnet
Thales Research and Technology, Palaiseau, France

W. Schmidt
Rosenberger Hochfrequenztechnik GmbH & Co. KG, Fridolfing, Germany

A. Shen · C. Calò
III-V Lab, Palaiseau, France

P. Smowton and T. Krauss (eds.), *Proceedings of the 26th European Conference on Integrated Optics*, Springer Proceedings in Physics 440,
https://doi.org/10.1007/978-3-032-29673-3_18

gap between theoretical circuit design and practical manufacturing for advanced PIC applications.

18.1 Introduction

The rapid advancement of photonic integrated circuit (PIC) technologies has revolutionized the field of photonics, increasing the growth of novel material platforms with enhanced performance. This progress has been further accelerated by the increasing demand for high-speed and energy-efficient computing driven by artificial intelligence (AI) and digital transformation [1]. While the primary markets for PICs remain telecommunications and data transmission, their integration into emerging fields, such as quantum technologies, optical sensing, digital health monitoring, and biomedical devices, demonstrates their versatility and importance [2]. PICs' scalability, compactness, and low power consumption make them particularly attractive for these applications.

Among the promising PIC platforms, thin film lithium niobate on insulator (LNOI or TFLN) has attracted significant attention due to its exceptional optical properties [3]. These include a high electro-optic (EO) coefficient, strong second- and third-order nonlinearities, and a broad transparency window from 350 to 5500 nm. The past few years have shown several crucial advancements in LNOI photonics, such as high-speed, low-voltage electro-optic modulators [4], Kerr and EO optical frequency combs [5, 6], and efficient wavelength conversion via periodically poled lithium niobate waveguides (PPLN) [7]. Despite these achievements, the commercial adoption of LNOI on a large scale remains limited, primarily due to the absence of standardized process design kits (PDKs) that are essential for design scalability. To bridge this gap, we demonstrate the simulation module library enabling the design and optimization of LNOI PICs established for an open PIC foundry supported by Swiss Center for Electronics and Microtechnology (CSEM) [8, 9].

18.2 PDK Library

Realistic PIC design requires engineers to go beyond detailed optical simulations and account for fabrication tolerances and technology constraints. From the early stages of circuit-level simulations, ensuring that the foundry can manufacture the proposed structures with high adequacy is essential. This means that the functional behaviour of the device must reflect all foundry restrictions, covering not only fabrication limitations but also validated geometry parameters provided by the foundry. To achieve this, all relevant design rules, process constraints, and functional building blocks should be compiled into a single, standardized library. Such a library enables users to accurately simulate and optimize complex photonic circuits while ensuring that the final design remains manufacturable. This approach bridges the gap between theoretical circuit design and practical fabrication process, facilitating the development of high-performance, foundry-compatible PICs.

One of the main challenges of implementing an LNOI PDK is the anisotropy of lithium niobate, which results in a change in the effective mode index depending on the direction of the propagated wave. As a result, the waveguide elements must be flexibly implemented to account for their paths. Furthermore, all optimized devices, such as MMIs, edge and grating couplers, etc., require adjustments to their geometries based on their orientation to reflect the desired performance. To simplify the design process, multiple devices have been restricted to placement only in a vertical or horizontal orientation (as shown in Fig. 1), ensuring that the signal propagates consistently in the y- or z-direction.

The simulation PDK library has been implemented for the schematic-driven layout-aware design methodology, utilizing a custom PDK framework as described in [10]. It has been adjusted to the CSEM LNOI platform [9] and implemented in the VPIcomponentMaker Photonic Circuits as an additional toolkit. It contains the developed building blocks (BBs) adapted to their realistic design performance. It includes all the required information to enable the PIC design for third-party users, including a seamless interface with the layout PDK developed in Nazca Design [10].

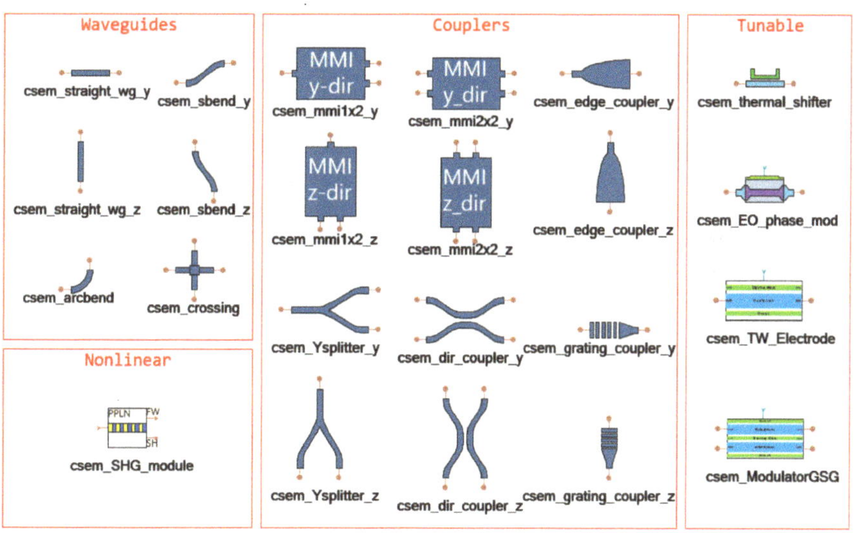

Fig. 1 Simulation PDK library components

18.3 Application Examples

To validate the PDK with realistic application cases using CSEM's LNOI foundry, three different end-user prototypes have been designed based on the PDK building block library. These prototypes will enable a performance assessment against state-of-the-art platforms from industrial foundries, demonstrating the advantages of LNOI over other PIC platforms. Furthermore, these demonstrators showcase the monolithic integration of various building blocks into more complex photonic circuits, serve as test designs to receive crucial feedback from end users during the early stages of technological development, and provide performance validation of the simulation PDK module library.

Rosenberger's prototype is a linearized electro-optic (EO) modulator. It takes advantage of low-loss, low V_π, and ultra-fast modulators to further demonstrate RF analog signal transmission over optical fibers. III-V Lab's application is in the telecom sector and high-speed optical transmission, leveraging fast-tunable cavities and optical switches to construct fast and widely tunable lasers. Thales' prototype is an IQ modulator operating as a frequency shifter, which can be applied to various applications, from quantum sensors to optical gyroscopes. The key LNOI functionalities for these prototypes are pure-phase shifters, frequency conversion (PPLN), and fast optical switches.

18.3.1 Linearized EO-Modulator

One of the primary difficulties in designing the linearized EO-modulator as a dual-parallel MZM (shown in Fig. 2) is the proper setting of the driving voltages, which are essential to achieve the desired modulator performance. Additionally, efficient power splitting and combining at the input and output couplers are required, as any imbalance can significantly decrease the overall performance. Signal propagation desynchronization in parallel parts might lead to a phase mismatch that could compromise functionality. Maximizing the efficiency of the electro-optic (EO) conversion remains a key challenge due to many local optima for the design parameters, which require careful balancing of multiple parameters to enhance the final system performance. Figure 2 shows the simulation schematic and example results illustrating the modulator's transfer function and intermodulation distortion.

A GDS of a test chip with a dual-parallel modulator is shown in Fig. 3.

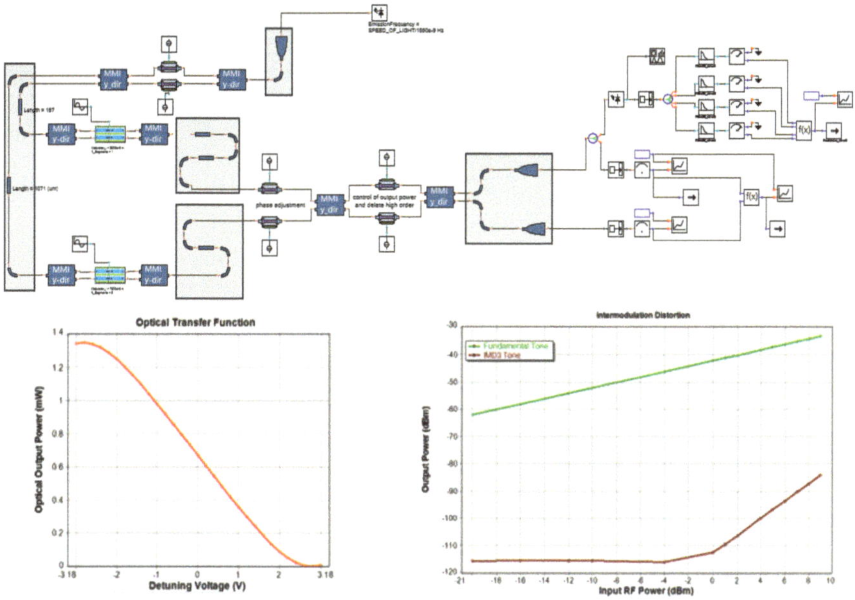

Fig. 2 Simulation schematic of the linearized EO-modulator with its optical transfer function and intermodulation distortion

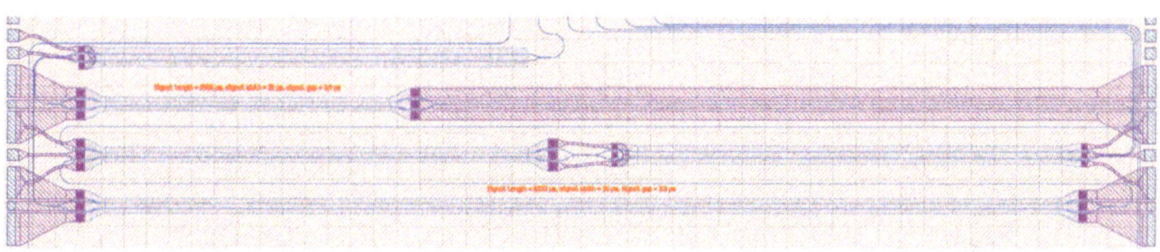

Fig. 3 GDS of the linearized EO modulator test chip to be manufactured by CSEM

18.4 High-Speed Widely Tunable Laser

The tunable laser design (shown in Fig. 4) presents several key challenges that impact its performance and functionality. The precise adjustment of ring resonator lengths specifies their free spectral range (FSR). These rings must also be further tuned using electrical DC signals to match the desired resonance frequency. This tuning mechanism requires very high stability to be suitable for most applications. Additionally, the output power is highly dependent on the semiconductor optical amplifier (SOA) facet reflectivity, which directly influences the signal output power. Another important consideration is the power splitting at the highlighted directional coupler, which defines the device's output. Specifically, a balanced 50/50 splitting directs the signal to output port 1 ("out1"), whereas any deviation from this ratio routes the remaining power to output port 2 ("out2"). The importance of these challenges is determined

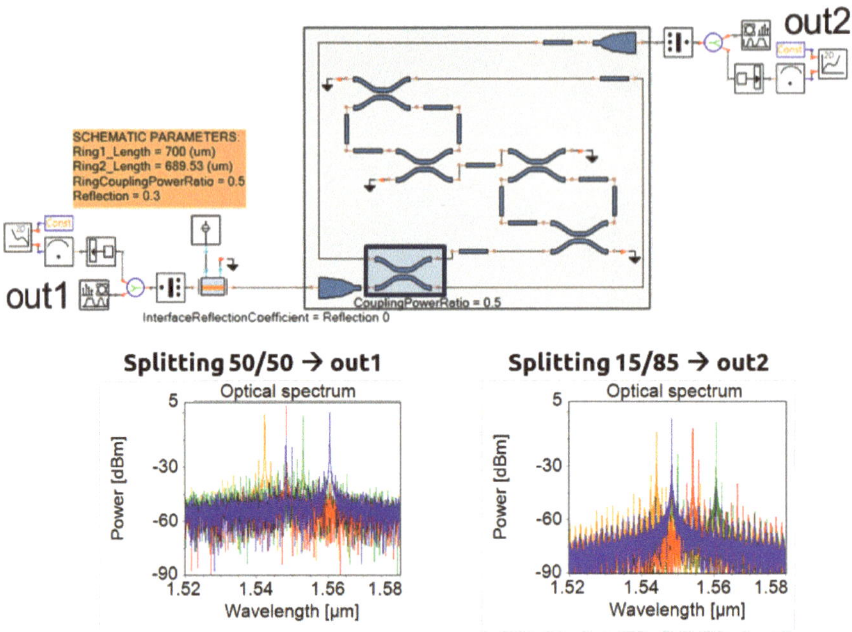

Fig. 4 Laser simulation schematic and its tuning characteristics for different splitting conditions of the marked directional couplers

by the desired laser performance, which varies according to the final application and system requirements. Figure 4 demonstrates the laser simulation schematic and its tuning characteristics for different splitting conditions.

18.4.1 Frequency Shifter

This frequency shifter design follows the architecture of a typical IQ modulator. One key challenge is the precise sideband generation, which requires a careful setting of thermo-optic shifters for DC biases and electro-optic (EO) phase shifters. At least 6.83 GHz frequency shifts with high stability and accuracy are required for quantum sensing applications based on cold rubidium atoms [11]. Furthermore, the metal crossings can introduce unwanted signal distortions and provide additional losses. Finally, adjusting the RF line lengths can minimize the occurrence of spurious wavelength peaks. Any mismatch between RF and optical signals could significantly degrade the system's functionality. Finally, PPLN is very sensitive to the fabrication tolerances and needs to be matched perfectly with the desired wavelength for high conversion efficiency. Addressing these challenges is essential to achieving a high-performance, reliable frequency shifter suitable for advanced quantum and photonic applications. Figure 5 demonstrates the simulation schematic and sideband generation for the fundamental and second harmonic signals. The modelling is a powerful tool to assess how the RF mismatch between the RF line lengths and the error in biasing might affect the spurious peaks.

This circuit was manufactured by CSEM with an additional optical filter stage, not represented in Fig. 6, and is now under packaging for testing. Figure 6 shows the chip GDS with its different IQ modulator, optical filter and PPLN conversion stages.

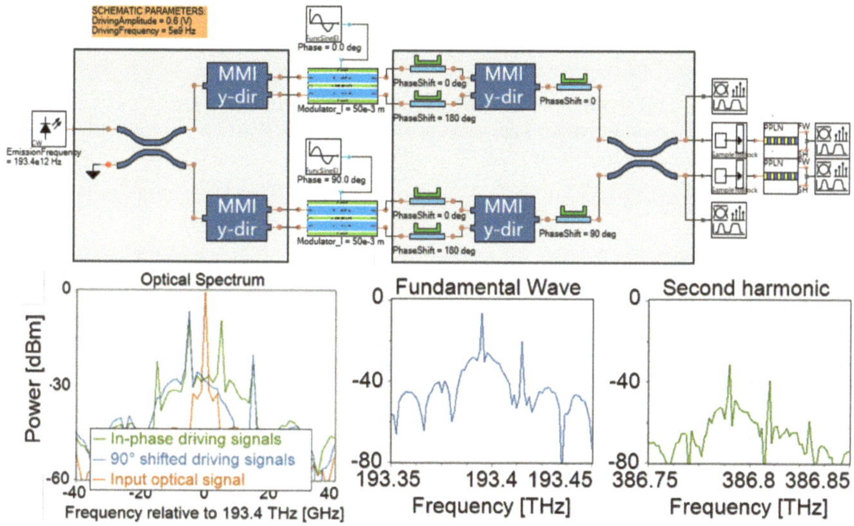

Fig. 5 Simulation schematic of the frequency shifter and results of optical spectra for different phases of the driving signals to generate sidebands of the fundamental and second harmonic signals

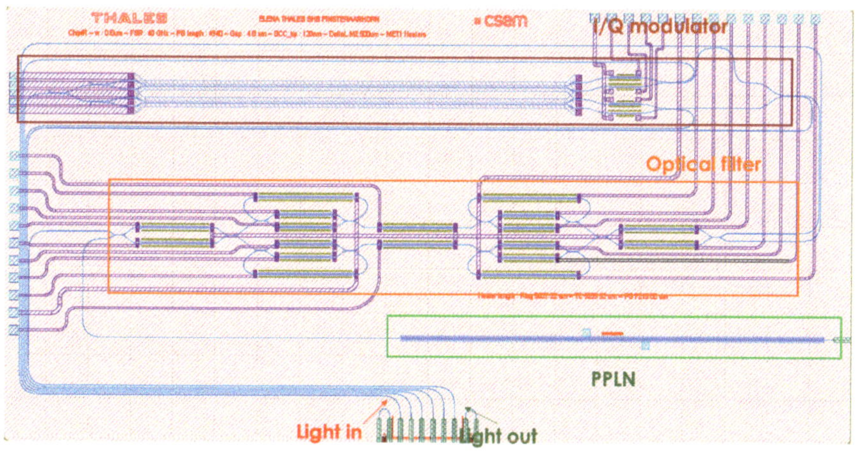

Fig. 6 GDS of the chip that is now under packaging for further testing

18.5 Conclusion

We report on our recent work performed in the ELENA project, which aims to develop a novel platform for PICs based on lithium niobate on insulator (LNOI). This versatile platform enables the design of numerous photonic applications for communications, sensing, and more. We focus on the design of a simulation PDK, which allows users to design application-specific PICs without having to develop essential building blocks. Furthermore, we demonstrate the PDK functionality with three different demonstrators: high-speed modulators, tunable lasers, and frequency shifters.

Acknowledgments This project has received funding from the European Union's Horizon 2020 research and innovation programme under grant agreement No 101016138.

References

1. Scholles, M. et al.: White Paper on Integrated Photonics, (White Paper) https://www.photonics21.org/download/ppp-services/photonics-downloads/White_Paper_on_Integrated_Photonics_-_EPoSS__Photonics21_-_ONLINEVERSION.pdf
2. Hänsel, A., Heck, M.J.R.: Opportunities for photonic integrated circuits in optical gas sensors. J. Phys. Photonics. **2**, 012002 (2020). https://doi.org/10.1088/2515-7647/ab6742
3. Boes, A., et al.: Lithium niobate photonics: Unlocking the electromagnetic spec-trum. Science. **379**, eabj4396 (2023). https://doi.org/10.1126/science.abj4396
4. Wang, C., et al.: Integrated lithium niobate electro-optic modulators operating at CMOS-compatible voltages. Nature. **562**(7725), 101–104 (2018). https://doi.org/10.1038/s41586-018-0551-y
5. Zhang, M., et al.: Broadband electro-optic frequency comb generation in a lithium niobate microring resonator. Nature. **568**(7752), 373–377 (2019). https://doi.org/10.1038/s41586-019-1008-7
6. Gong, Z., et al.: Soliton microcomb generation at 2 μm in z-cut lithium niobate microring resonators. Opt. Lett. **44**(12), 3182–3185 (2019). https://doi.org/10.1364/OL.44.003182
7. Wang, C., et al.: Second harmonic generation in nano-structured thin-film lithium niobate waveguides. Opt. Express. **25**(6), 6963 (2017). https://doi.org/10.1364/OE.25.006963
8. Sattari, H., et al.: Standardized TFLN photonic integrated circuits platform. In: Witzens, J., Poon, J., Zimmermann, L., Freude, W. (eds.) The 25th European Conference on Integrated Optics. ECIO 2024. Springer Proceedings in Physics, vol. 402. Springer, Cham (2024). https://doi.org/10.1007/978-3-031-63378-2_15
9. CSEM.: Foundry services for TFLN lithium niobate integrated photonics. Retrieved from https://www.csem.ch/en/tailored-services/tfln-foundry-services/
10. Polatynski, A., Savitski, S., Maloney, C., Richter, A.: The best of both worlds: Design freedom and IP protection. In: PIC Magazine, Issue, vol. 1, pp. 12–14 (2024) https://picmagazine.net/article/118912/The_Best_Of_Both_Worlds_Design_Freedom_And_IP_Protection/feature
11. Steck, D.A.: Rubidium 87 D Line Data. https://steck.us/alkalidata/rubidium 87numbers.pdf

Chapter 19
Novel Contra-Directional Coupler Devices Based on Waveguide Width Chirp

Patrícia Carvalho, Joana Tatá, Inês Venâncio, João Santos, Paulo Duarte, Carla Rodrigues, Francisco Rodrigues, and António Teixeira

Abstract The ever-growing requirements of optical communications, from higher data transmission speeds and larger capacities to lower costs, have prompted the development of PIC-based transceivers integrating, among others, optical filters for Wavelength Division Multiplexing (WDM). Considering the lack of mature O-band wavelength filters, we have simulated and characterized contra-directional couplers based on waveguide width chirping to tackle the typical low rejection and weak coupling efficiency.

19.1 Introduction

The increasing complexity of Passive Optical Networks (PON) standards has instigated the development of photonic integrated solutions to achieve low-cost, high performance pluggables [1, 2]. In this context, optical filters are key building blocks for wavelength multiplexing and demultiplexing operations, that must meet rigorous criteria in terms of bandwidth, isolation, and footprint. Various approaches have been considered for this purpose, from Arrayed Waveguide Gratings (AWG) to Mach-Zender interferometers (MZI) or ring resonators. However, these devices are either limited in terms of free spectral range (FSR) or in terms of footprint. Alternatively, Bragg based structures, namely contra-directional couplers (contra-DC),

P. Carvalho (✉) · J. Tatá · I. Venâncio · J. Santos · P. Duarte · C. Rodrigues · F. Rodrigues · A. Teixeira
PICadvanced, S.A., PCI—Creative Science Park, Ílhavo, Portugal
e-mail: patricia.carvalho@picadvanced.com

J. Santos · A. Teixeira
Instituto de Telecomunicações (IT), Universidade de Aveiro, Aveiro, Portugal

Departamento de Electrónica, Telecomunicações e Informática (DETI), Universidade de Aveiro, Aveiro, Portugal

P. Smowton and T. Krauss (eds.), *Proceedings of the 26th European Conference on Integrated Optics*, Springer Proceedings in Physics 440,
https://doi.org/10.1007/978-3-032-29673-3_19

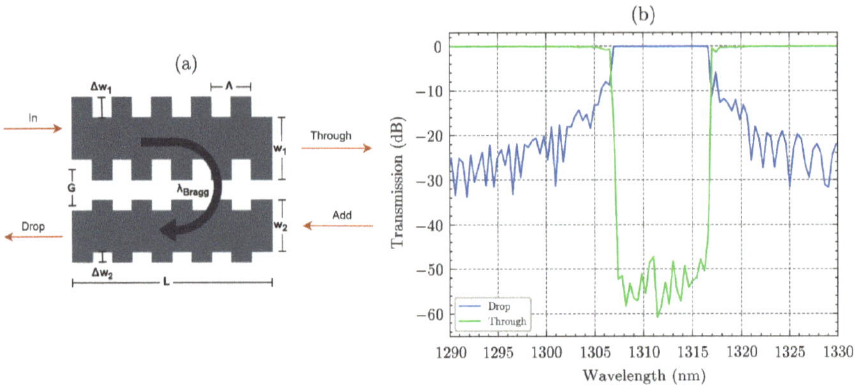

Fig. 19.1 **a** Schematic representation of a contra-directional coupler, highlighting the *in*, *drop*, *through*, and *add* ports. **b** *Drop* and *through* responses of a simulated contra-DC

show promise for a compact device with flat-top response, low insertion loss and low crosstalk.

A contra-DC is a four-port device acting as an add-drop filter, that is composed of two parallel anti-symmetric Bragg gratings with different widths, as shown in Fig. 19.1a, at a coupling distance that allows some optical mode overlap. Light travelling along the first waveguide partially reflects at each corrugation and constructive interference occurs for wavelengths satisfying the Bragg condition. The reflected wavelength is coupled into the second waveguide and exits at the drop port, while the remaining light exits at the through port [3, 4].

The response of a contra-DC, shown in Fig. 19.1b, is severely influenced by its geometry, with the reflected wavelength being determined by the corrugation period (Λ) and the width of the waveguides ($w_{1,2}$). Parameters such as corrugation width ($\Delta w_{1,2}$) and gap (G) can influence the coupling coefficient and thus the filter's bandwidth. Hence, the filter's response can be fine-tuned by implementing geometry variations. For example, waveguide width chirping, i.e., the gradual variation of the width of the waveguides, increases bandwidth, by enabling multiple wavelengths to meet the Bragg condition through the varying effective index along the propagation axis. Two main chirping approaches can be taken: symmetric waveguide width variation in both vertical directions, and asymmetric variation in the outer vertical direction, as shown in Fig. 19.2a. Chirp distribution, shown in Fig. 19.2b, can be either linear, where each grating section with different widths has the same number of corrugations, or non-linear, where the number of corrugations varies for different widths.

To meet the demands of optical communications, channel bandwidths beyond the C-band must be exploited, and so it is crucial to pursue solutions for the O-band and mature wavelength division multiplexing filters are necessary. Since solutions compatible with commercial fabrication with high rejection and strong coupling are still scarce, we propose novel contra-DC configurations with linear asymmetric and

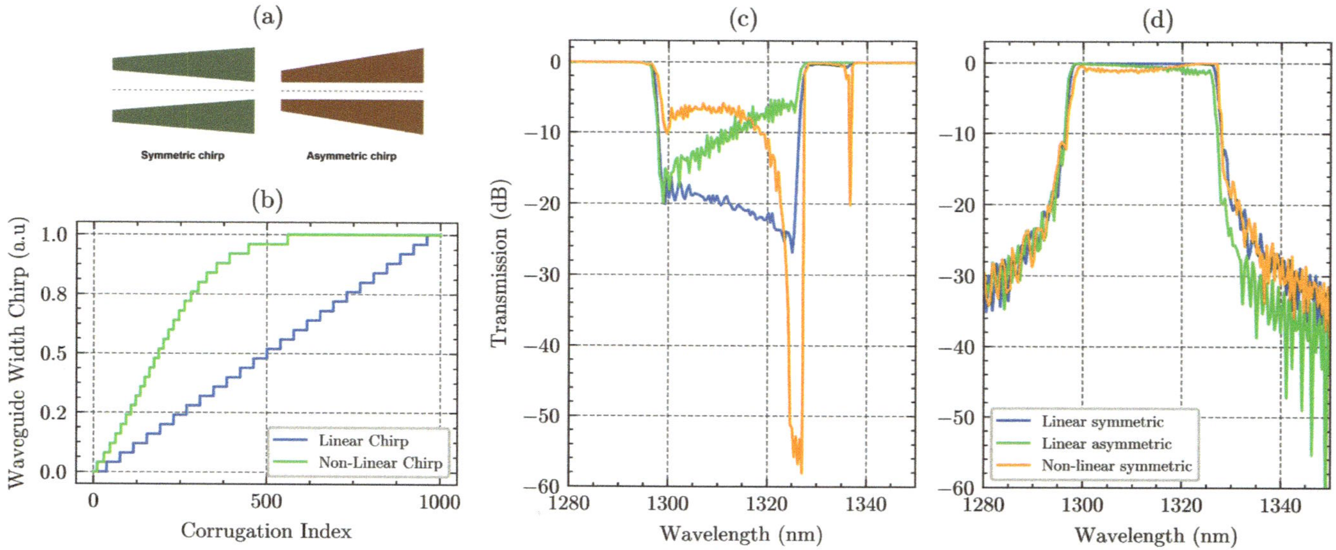

Fig. 19.2 **a** Chirped waveguide symmetry. **b** Linearity of chirp distributions. **c** and **d** Simulated *through* and *drop* responses, respectively, of the different contra-DC configurations

non-linear symmetric waveguide width chirping. These are compared to a standard linear symmetric chirped contra-DC.

19.2 Methodology

19.2.1 Design and Simulation

To understand the effect of waveguide chirp on the performance of contra-DCs, three different configurations were designed and simulated: linear asymmetric, non-linear asymmetric, and linear symmetric.

Simulation was performed using a modified version of the Lumerical EigenMode Expansion simulation scripts of the Contra-DC structure provided by the SIEPIC group [5]. Alterations were made to the original scripts to include the effect of the distinct types of waveguide width chirps. Multiple cell groups were included, each corresponding to a different chirp "step", with a periodicity equal to the total number of periods in the structure divided by the total number of steps. These steps were highly influenced by the design and fabrication limitations: a 2 nm step of waveguide width was necessary to implement the chirp feature in the design with maximum control.

19.2.2 Experimental Characterization of Contra-DCs

The simulated contra-DCs were fabricated in a 220 nm SOI platform, through 193 nm DUV lithography in a dedicated prototyping run. *Drop* and *through* responses of the devices were characterized through grating couplers, using a broadband light source as input and an optical spectrum analyzer (OSA) to measure the optical spectra at the outputs as shown in Fig. 19.3a.

19.3 Results and Discussion

The simulated *through* and *drop* responses of the different contra-DCs are shown in Fig. 19.2c and d. The *through* responses show varying rejection levels across the devices. The linear symmetric chirped contra-DC (blue) has 20 to 25 dB rejection increasing with wavelength, and the non-linear symmetric device (orange) shows a similar behavior, but with higher rejection for higher wavelengths. On the other hand, the rejection of the linear asymmetric contra-DC (green) decreases with increasing wavelengths.

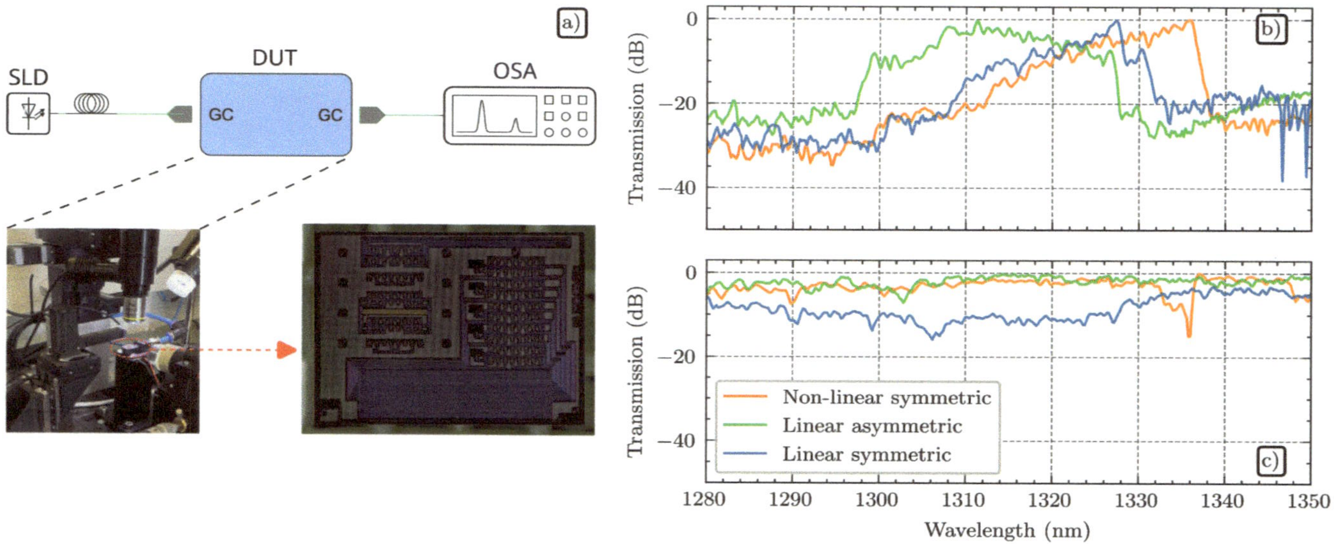

Fig. 19.3 **a** Experimental setup for device characterization. **b** and **c** *Drop* and *through* responses of the non-linear symmetric (orange), linear asymmetric (green), and linear symmetric (blue) contra-DCs

Figure 19.3b and c shows the experimental *through* and *drop* responses of the fabricated contra-DCs, that present a wavelength shift from simulation due to fabrication deviations [6, 7]. Experimental measurements of linear symmetric and linear asymmetric contra-DCs show no rejection in the *through* responses, while the non-linear asymmetric device presents 10 to 15 dB rejection. However, it is unclear whether this improvement is due to the non-linear nature of the chirp or because the last chirp step has met the minimal number of periods necessary for a measurable rejection. Regarding the measured *drop* responses, both the non-linear symmetric and linear symmetric devices translate the increased rejection for higher wavelengths that is observed in the through responses.

These results highlight some design considerations affecting the coupling efficiency of the devices. Waveguide gap width is crucial, especially for symmetrically chirped devices, where the minimal gap is constrained by fabrication guidelines. For this case, the varying gap along the contra-DC length leads to increasing coupling with increasing wavelengths that are filtered by higher chirp values and experience the minimal spacing. Conversely, asymmetric chirped devices maintain consistent spacing but experience decreased coupling at higher wavelengths due to stronger mode confinement in wider waveguides.

19.4 Conclusions

In this paper, different contra-DC configurations based on waveguide width chirping were investigated for WDM in the O-band.

The performance of linear symmetric, linear asymmetric, and non-linear symmetric chirped devices was simulated and experimentally assessed, towards the development of a device with high rejection and strong coupling performance. While simulation results reveal higher rejection ratios for the linear symmetric case, experimental results show discrepancies: only the non-linear asymmetric device achieved meaningful rejection values. Design constraints were also taken into account, providing insight into future contra-DC optimization focusing on geometry parameters such as device length.

References

1. Teixeira, A.: Photonic integrated circuits for NG-PON2 ONU transceivers. In: Optical Fiber Communication Conference (OFC) 2019, OSA Technical Digest. Optica Publishing Group (2019) paper M4G.4
2. Pinho, C., Rodrigues, F., Tavares, A.M., Rodrigues, C., Rodrigues, C.E., Teixeira, A.: Photonic integrated circuits for NGPON2 ONU transceivers (invited). Appl. Sci. **10**, 4024 (2020). https://doi.org/10.3390/app10114024
3. Cauchon,J.: Silicon Photonic Bragg-based Devices: Hardware and Software, PhD Thesis. Université Laval (2021)

4. Shi, W., Wang, X., Lin, C., Yun, H., Liu, Y., Baehr-Jones, T., Hochberg, M., Jaeger, N.A.F., Chrostowski, L.: Silicon photonic grating-assisted, contra-directional couplers. Opt. Express. **21**, 3633–3650 (2013)
5. Hammood, M.: SIEPIC Contra-DC Simulator 2023. https: //github.com/SiEPIC/contraDC
6. Venâncio, I., Tátá, J., Carvalhais, M.J., Santos, J., Teixeira, A.: Prediction through simulation-based corrective models of contra-directional couplers' experimental results. Photonics. **11**, 774 (2024)
7. Venâncio, I., Tátá, J., Santos, J., Duarte, P., Rodrigues, C., Teixeira, A.: Impact of process variations on the performance of wavelength-dependent devices. In: Young Kim, K. (ed.) Optical Waveguide Technology and and Applications. Intech Open (2024)

Chapter 20
Characterization and Prediction of Wafer-Level Performance Variations on Silicon Photonic Devices

Inês Venâncio, João Santos, Patrícia Carvalho, Paulo Duarte, Joana Tátá, Carla Rodrigues, Francisco Rodrigues, and António Teixeira

Abstract Photonic integrated circuits have been seeing a steady surge in number of devices, density and overall complexity. As such, measuring performance variations in PICs that can have up to thousands of devices can become extremely costly and time-consuming. This paper presents findings that indicate a strong performance correlation between different devices within a PIC across multiple wafers. Based on these results, a model to predict these wafer-level spectral shifts in passive devices using a single structure was developed which presented similar results to the measured data.

20.1 Introduction

The possibility of using photonic integrated circuits (PICs) in passive optical networks has been gaining traction due to the ever-increasing requirements to enlarge the bandwidth of these systems [1] given the massive traffic growth observed in the last decades.

Silicon photonics is one of the most popular and widely used platforms in photonic integration, given its low fabrication cost and process maturity owing to the import of CMOS fabrication tools. However, despite the constant efforts to optimize fabrication techniques and lithography resolution in an attempt to reduce the well-known issue of wafer variability in silicon photonics, this problem remains a major reason why the performance of passive photonic devices is still unreliable, since it is very

I. Venâncio (✉) · J. Santos · P. Carvalho · P. Duarte · J. Tátá · C. Rodrigues · F. Rodrigues · A. Teixeira
PICadvanced S.A., Ílhavo, Portugal
e-mail: ines.venancio@picadvanced.com

J. Santos · A. Teixeira
Instituto de Telecomunicações, Aveiro, Portugal

Universidade de Aveiro, Aveiro, Portugal

P. Smowton and T. Krauss (eds.), *Proceedings of the 26th European Conference on Integrated Optics*, Springer Proceedings in Physics 440,
https://doi.org/10.1007/978-3-032-29673-3_20

strongly dependent on fabrication conditions, such as thickness and waveguide width variations [2].

Given the surge in the number of structures and overall complexity of PICs, understanding and quantifying the impact of fabrication variations in each individual device within a PIC on a wafer level can become extremely time-consuming and expensive due to the necessity of mass testing and setup complexity required.

This paper analyzes passive photonic devices at wafer level, identifying variations across structures and establishing a correlation between their spectral shifts, enabling performance prediction of multiple devices based on a single structure.

20.2 Test Structures and Methodology

Three devices have been chosen for this study based on their popularity as optical filters in integrated photonics as well as their difference in geometry complexities: an optical ring resonator (ORR), a Bragg grating and a contra-directional coupler (CDC). The design of the ORR was based on the work of Lu et al. [3] while the Bragg grating and CDC were designed in-house. The devices were fabricated in 2 equal 220-nm SOI wafers in a dedicated prototyping run. Each wafer has 54 identical reticles, each reticle containing the 3 structures in study.

These structures were characterized with a broadband super-luminescent optical light source as input, a fiber array to inject and receive radiation to and from the PIC via grating couplers and an optical spectrum analyzer at the output to acquire the spectra of each device. Based on the results of the first wafer, a second one was also tested to validate the prediction model developed.

20.3 Results and Discussion

The central wavelength shift observed for the ORRs, Bragg gratings and CDCs throughout the 54 reticles of the first wafer can be observed in Fig. 20.1. The effects of process variations on the performance of all 3 devices are clearly visible with results spanning across a wavelength interval of more than 20 nm and varying erratically between reticles.

Despite the large shifts present on all devices, it is possible to identify a similar trend among the different structures, implying a direct correlation in the spectral shift of the components. Hence, the variability is caused by a systematic effect in the fabrication process of the wafer and, therefore, affects all structures on a PIC in a similar manner, causing similar shifts. Thus, by testing one structure across the whole wafer, it should be possible to infer the shift of any structure without needing to experimentally measure it.

To validate this hipothesis, a second wafer was tested and the wavelength shift of the structures was predicted using the results of the first wafer. For this model,

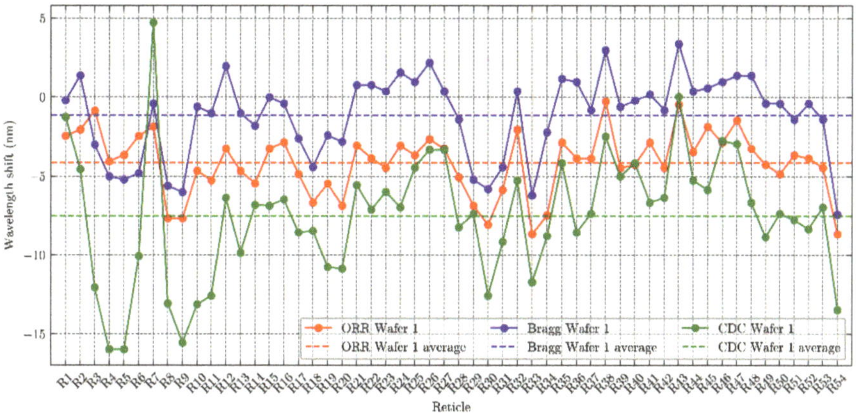

Fig. 20.1 Central wavelength shift observed for the 3 structures under testing (ORR, Bragg grating and ORR) across a wafer

one of the devices was used as a reference to map the wavelength shift difference of the remaining structures in wafer 1. This difference was then applied to the results obtained in wafer 2, using the same reference component, to understand if the prediction matched the measured data.

Figure 20.2 shows the density distribution of the experimental central wavelength shift of the Bragg grating for the 2 wafers tested. It is visible that, despite the slightly larger overall shift seen in wafer 2, the distribution is similar to the one observed in wafer 1, evidenced by the similar gaussian fits of the two datasets.

The shift difference observed between devices in wafer 1 were then applied to the structures of wafer 2 with the Bragg grating as reference to predict the shift of the ORRs and CDCs throughout the whole wafer. The results of this model and measured data of wafer 2 can be seen in Fig. 20.3, where it is visible that the predicted trend closely follows the experimental data across the whole wafer. The model results present an average difference to measured results of -0.6 ± 1.6 nm for the ORR and $- 1.23 \pm 1.8$ nm for the CDC. From these results, we infer that the presented model can successfully predict the devices' performance across multiple wafers based on the characterization of a single structure.

It is worth noting that any of the 3 devices could have been chosen as the reference to this model, however, the more complex nature of the CDC's geometry makes it more susceptible to fabrication variations [4] and is therefore not ideal since it has a tendency to present larger and more random shifts than the ORR and Bragg grating, also evidenced by the larger error in the model.

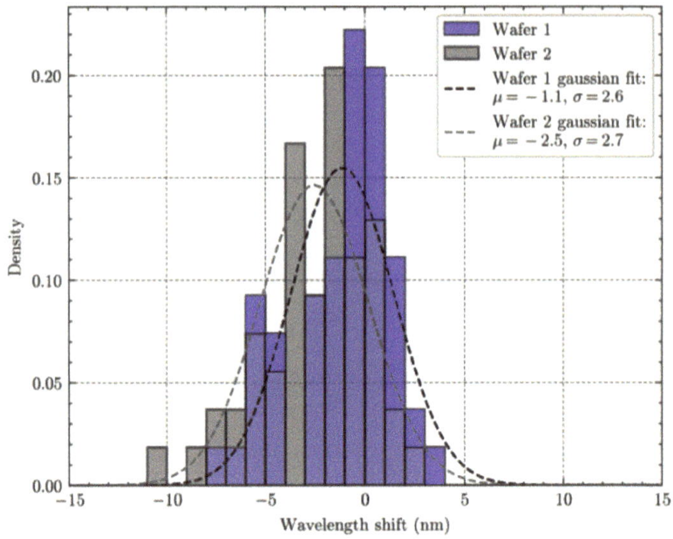

Fig. 20.2 Density distribution of the central wavelength shift observed for the Bragg grating on the 2 wafers

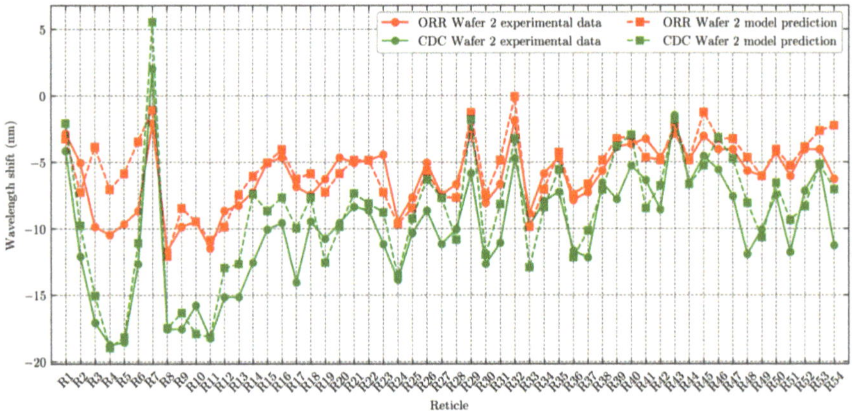

Fig. 20.3 Experimental data and model prediction results for the central wavelength shift of ORRs and CDCs using the Bragg grating as reference

20.4 Conclusions

This paper investigated the performance variations observed in passive photonic devices which stem from fabrication variability. Through this study, a direct correlation in the central wavelength shift observed in the different structures, namely optical ring resonators, Bragg gratings and contra-directional couplers, was proved. This correlation was used to successfully predict the shift of devices on a different wafer with modelled results shifting from measured data on average only 0.6 nm for the ORR and 1.23 nm for the CDCs using the Bragg grating's results to predict it. The correlation found thus presents the potential to save considerable time and resources on characterization efforts for future wafers, proportionally to the number of devices in a given PIC.

References

1. Rodrigues, F., Rodrigues, C., Santos, J., et al.: Photonic integrated circuits for passive optical networks: outlook and case study of integrated quasi-coherent receiver. Photonics. **10**, 182 (2023). https://doi.org/10.3390/photonics10020182
2. Xing, Y., Dong, J., Khan, U., Bogaerts, W.: Capturing the effects of spatial process variations in silicon photonic circuits. ACS Photonics, 928–944 (2022). https://doi.org/10.1021/acsphotonics.2c01194
3. Lu, Z., Jhoja, J., Klein, J., et al.: Performance prediction for silicon photonics integrated circuits with layout-dependent correlated manufacturing variability. Opt. Express. **25**, 9712 (2017). https://doi.org/10.1364/OE.25.009712
4. Venâncio, I., Tátá, J., Santos, J., et al.: Impact of process variations on the performance of wavelength-dependent devices. In: Optical Waveguide Technology and and Applications. Intech Open (2024). https://doi.org/10.5772/intechopen.114847

Chapter 21
Computationally Efficient Simulation of Multimode Interference Reflectors in the Frequency Domain

George M. Jandu and Peter M. Smowton

Abstract The optimum design strategy for multimode interference reflectors is investigated by comparing the accuracy of and the time taken for several methods to address an exemplar, practical problem of optimising device length for a given wavelength in the O-band. Previously, three-part simulations based on the finite difference time domain method combined with either a fast uni-directional or bi-directional approach have been employed. Here we present a new formulation of an equivalent geometry approach based on the eigenmode expansion technique, and demonstrate that it can be used to find the optimum length to within 0.5 μm, over 130 times quicker than a typical GPU accelerated three-part approach. This suggests that the optimal strategy is to use the equivalent geometry approach to efficiently map the parameter space, and then an accurate three-part simulation to finalise the design.

21.1 Motivation

Multimode interference reflectors (MMIRs) are versatile devices with applications as broadband reflectors, power splitters [1], and temperature insensitive laser devices [2]. The optical simulation of MMIRs is complicated by the inherently omni-directional propagation of light between the angled mirrors; this makes the fast uni- and bi-directional techniques typically used in computational photonics unsuitable. Omni-directional techniques like the finite difference time domain (FDTD) method are computationally intensive, meaning that full device simulations are unfeasible on personal computers. First introduced by Kleijn et al. in 2013, two classes of approach have previously been used to circumvent these issues: the equivalent geometry approach and three-part simulations [3].

G. M. Jandu (✉) · P. M. Smowton
Cardiff University, Cardiff, UK
e-mail: JanduGM@cardiff.ac.uk

P. Smowton and T. Krauss (eds.), *Proceedings of the 26th European Conference on Integrated Optics*, Springer Proceedings in Physics 440,
https://doi.org/10.1007/978-3-032-29673-3_21

As they are currently formulated, neither approach is well suited to efficient parameter scans, and therefore efficient device optimisation. In this work, we seek to address this issue by comparing the results (and the timescales required to achieve them) of several reformulations and alternative techniques for the simulation of MMIR reflectivity. We focus in particular on the practical problem of optimising the length of the reflector for a given wavelength.

21.2 Simulation

21.2.1 The Equivalent Geometry Approach

For the idealised case of perfectly vertical, perfectly reflecting side walls, and no rounding of the vertex where the two 45° mirrors meet, it is possible to calculate the reflected field in the straight waveguide portions of the device directly from the forward propagating field incident on the reflector [3]. Under these assumptions, the effect of the angled mirrors on the forward propagating field is threefold: the direction of propagation is reversed, the shape of the field is mirrored along the plane of symmetry, and there will be some mode dependent phase shift associated with the reflections.

Neglecting the phase shifts for now, and denoting the forward and backward travelling waves as $\mathscr{F}^+$ and $\mathscr{F}^-$ respectively, if we take the coordinate system to be defined as in Fig. 21.1, then the backward propagating field at any point in the device can be written in terms of the forward propagating field that would be present if the multimode waveguide was of semi-infinite extent in the z-direction:

$$\mathcal{F}^-(x, y, z) = \mathcal{F}^+(-x, y, 2L - z), \ \forall x \forall y \forall z \in \{z : 0 \leq z \leq L\}. \tag{1}$$

Here L is the total length of the MMIR excluding the input waveguide. The forward propagating field can be calculated using uni-directional or bi-directional techniques. Kleijn et al. employed the beam propagation method (BPM), but with this approach the entire simulation must be repeated when a parameter is changed. We will now present a formulation based on the eigenmode expansion technique (EME) which lends itself well to parameter scans.

If the eigenvectors of the cross-section in the x-y plane of a device are calculated using the finite difference method, then the transverse fields at any point along the z-axis (assuming uniformity in the z-direction) are given by [4]:

$$\psi(z) = \begin{bmatrix} \boldsymbol{e}_x(z) \\ \boldsymbol{e}_y(z) \\ \boldsymbol{h}_x(z) \\ \boldsymbol{h}_y(z) \end{bmatrix} = \begin{bmatrix} \boldsymbol{W} & \boldsymbol{W} \\ \boldsymbol{V} & -\boldsymbol{V} \end{bmatrix} \begin{bmatrix} \exp(-\boldsymbol{\lambda} k_0 z) & 0 \\ 0 & \exp(\boldsymbol{\lambda} k_0 z) \end{bmatrix} \begin{bmatrix} \boldsymbol{c}^+ \\ \boldsymbol{c}^- \end{bmatrix}. \tag{2}$$

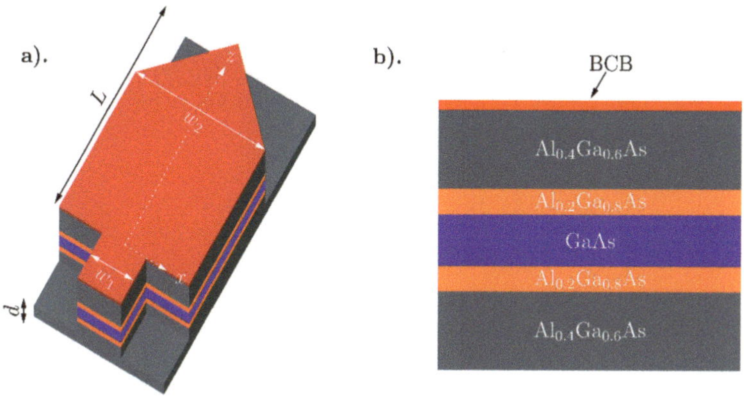

Fig. 21.1 The multimode interference reflector simulated in this work. **a**). Schema showing the coordinate system and important parameters. Here $w_1 = 3$ μm, $w_2 = 6$ μm, and $d = 0.35$ μm. **b**). The layer structure of the device

Here, $\boldsymbol{c}^+$ and $\boldsymbol{c}^-$ are vectors containing the mode coefficients for the forward and backward propagating fields respectively, $\boldsymbol{W}$ is a matrix where each column is an eigenvector for the electric field components, $\boldsymbol{V}$ is a matrix where each column is an eigenvector for the magnetic field components (which here are normalised by a factor of the impedance of free space), k_0 is the free space wavevector of the fields, and $\boldsymbol{\lambda}$ is a diagonal matrix containing the effective index of each mode multiplied by the imaginary number i.Denoting quantities in the input and multimode waveguides with subscript i and m respectively, the boundary conditions at the interface between the input and multimode waveguide can be written:

$$\begin{bmatrix} \boldsymbol{W}_\mathrm{i} & \boldsymbol{W}_\mathrm{i} \\ \boldsymbol{V}_\mathrm{i} & -\boldsymbol{V}_\mathrm{i} \end{bmatrix} \begin{bmatrix} \boldsymbol{c}_\mathrm{i}^+ \\ \boldsymbol{c}_\mathrm{i}^- \end{bmatrix} = \begin{bmatrix} \boldsymbol{W}_\mathrm{m} & \boldsymbol{W}_\mathrm{m} \\ \boldsymbol{V}_\mathrm{m} & -\boldsymbol{V}_\mathrm{m} \end{bmatrix} \begin{bmatrix} \boldsymbol{c}_\mathrm{m}^+ \\ \boldsymbol{c}_\mathrm{m}^- \end{bmatrix}. \tag{3}$$

For the semi-infinite multimode waveguide used in this analysis, there are no reflections, so $\boldsymbol{c}_\mathrm{m}^+ = 0$, and Eqs. 2 and 3 can be combined to obtain the fields at $z = 2L$:

$$\psi(2L) = \begin{bmatrix} \boldsymbol{W}_\mathrm{m} \exp(-2\boldsymbol{\lambda}_\mathrm{m} k_0 L) \boldsymbol{A}_\mathrm{im} \\ \boldsymbol{V}_\mathrm{m} \exp(-2\boldsymbol{\lambda}_\mathrm{m} k_0 L) \boldsymbol{A}_\mathrm{im} \end{bmatrix}, \tag{4}$$

$$\boldsymbol{A}_\mathrm{im} = \boldsymbol{W}_\mathrm{m}^{-1} \boldsymbol{W}_\mathrm{i} \Big(1 + (\boldsymbol{W}_\mathrm{m}^{-1} \boldsymbol{W}_\mathrm{i} + \boldsymbol{V}_\mathrm{m}^{-1} \boldsymbol{V}_\mathrm{i})^{-1} (\boldsymbol{V}_\mathrm{m}^{-1} \boldsymbol{V}_\mathrm{i} - \boldsymbol{W}_\mathrm{m}^{-1} \boldsymbol{W}_\mathrm{i}) \Big) \boldsymbol{c}_\mathrm{i}^+. \tag{5}$$

In this formulation, the eigenmode expansion equivalent geometry approach (EME-EGA) to calculating the backward propagating wave at $z = 0$ is straightforward:

1. Calculate the modes for both the input and multimode waveguides
2. For a given set of input mode coefficients, calculate the field at $z = 2L$ using Eq. 4

3. Mirror the field in the plane of symmetry of the device

The majority of the computation time elapses during step 1. Fortunately, to determine the reflectivity for several lengths, only steps 2 and 3 must be repeated, enabling efficient device optimisation.

21.2.2 Three-Part Simulations

To include losses and phase shifts due to the reflections, omni-directional techniques must be incorporated into the simulation. To leverage both the speed of uni-directional techniques and the accuracy of omni-directional techniques, hybrid simulations are often employed. Uni- or bi-directional methods are used to propagate the light through the straight waveguide sections, and the resulting field is launched into an omni-directional solver for the angled mirror section. The reflected field can then be loaded back into the uni- or bi-directional solver to propagate the light back to the input port.

In recent publications, both EME [5] and BPM [1] have been used in the straight waveguide sections and FDTD is typically used in the angled mirror section. A significant proportion of the computation time is spent waiting for the FDTD simulation to reach the steady state. Consequently, a significant amount of time can be saved by using the finite difference frequency domain (FDFD) method to calculate the steady state solution directly.

Incorporating sources into frequency domain simulations amounts to performing a matrix division of the form [6]:

$$f = \boldsymbol{A}^{-1} b, \tag{6}$$

Where $\boldsymbol{f}$ is the steady state field vector, $\boldsymbol{b}$ is a vector calculated using the well known 'QAAQ' equation, and $\boldsymbol{A}$ is a matrix constructed from the material tensors and derivative matrices on the Yee grid [7]. The matrices involved in these calculations are extremely large for 3D simulations and Eq. 6 must often be solved iteratively. Modern computers can accelerate this process by performing the calculation on the graphics processor unit (GPU), reducing computation time by factors as large as 20 [8]. To implement a hybrid simulation, Eq. 4 can be used to propagate the fields along the majority of the straight waveguide section, Eq. 6 can be solved via the GPU, and the reflected field can be back propagated again using Eq. 4.

This three-part approach improves the accuracy of the calculation, but is poorly suited to parameter scans, as the entire simulation must be repeated for any change. In principle, an effective scattering matrix could be calculated for the angled mirror section enabling efficient length scans, but this will only save time when the number of lengths required is greater than the number of modes. Accordingly, three-part simulations are best suited to cases where few modes are required, or the number of cells required is sufficiently small that the total computation time per simulation is low.

21.2.3 Two Dimensional Approximations

It is possible to create two dimensional approximations to three dimensional problems with the effective index method (EIM). When modelling multimode devices, using the EIM is complicated by the choice of which mode to use for the effective index. Using the fundamental mode of the input waveguide leads to incorrect predictions for the beat length, altering the transmission and reflection of the device. For multimode interference couplers, it has been shown that this issue can be addressed by varying the off-rib effective index [9].

This method cannot be applied directly to MMIRs as changing the ratio of the refractive indices will affect the reflection at the interfaces. In this work, we use a modified version of this improved EIM (IEIM), in which we scale both effective indices by the same amount to correct the beat length whilst preserving the reflectivity.

It is possible to halve the computational domain through the use of perfect magnetic conductor (PMC) and perfect electric conductor (PEC) boundary conditions along the plane of symmetry of the device, when injecting transverse electric (TE) and transverse magnetic (TM) modes respectively [10]. 2D representations in conjunction with the correct boundary conditions reduce the computational domain sufficiently to enable full device simulations with either the FDTD or FDFD method.

21.3 Results and Discussion

The results of simulating the device shown in Fig. 21.1 with each of the methods described above are plotted in Fig. 21.2. There is a clear discrepancy in the magnitude of the reflectivity predicted by the three methods; this is likely a result of the 2D FDFD approximation incorrectly predicting (and the EME-EGA neglecting) the mirror losses. However, all three approaches predict the same optimum length to within 0.5 μm; indicating that reflection induced phase shifts are of little importance.

The times taken to run each simulation (measured using MATLAB's tic and toc functions) are collated in Table 21.1. The reformulation of the EGA presented here performs the length scan over 130 times faster than the GPU accelerated hybrid FDFD approach, which will significantly reduce the time required for the initial coarse optimisation of device parameters. This becomes more important for the full device optimisation when there are more free parameters; for example, while the width of the input waveguide is often fixed by external factors, the width of the multimode section must be selected. Each time the width is altered, the modes must be recalculated, but the length can then be scanned efficiently as described above, meaning that the use of the EME-EGA over three-part simulations allows the

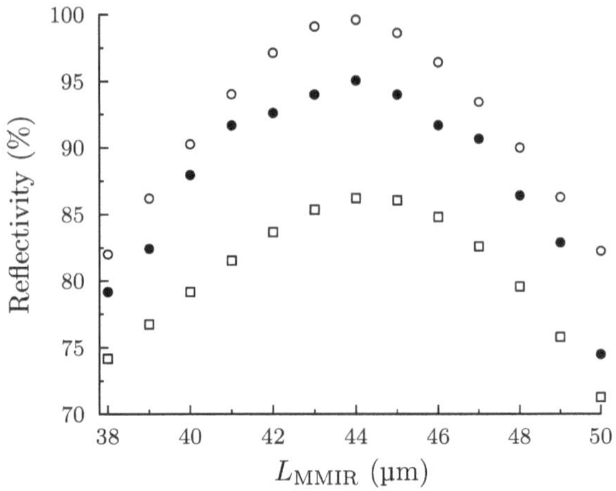

Fig. 21.2 The simulated reflectivity of a multimode interference reflector when injected with the fundamental mode at a wavelength of 1.36 μm, with the EME-EGA (open circles), the 2D FDFD method (filled circles), and the three-part EME/FDFD approach (open squares)

Table 21.1 Comparison of the run times for different simulation methods for multimode interference reflectors. These simulations were run on a computer with a NVIDIA GeForce RTX4090 GPU and an Intel Core i9 14900HX processor

Method	Average single run time (s)	Length scan time (s)
EME-EGA	10.6	12.3
2D FDFD	11.5	149.5
EME-FDFD hybrid	125.6	1632.8

parameter space to be mapped in a matter of minutes rather than hours. Accordingly, the optimal design strategy is to use the EME-EGA to quickly find an initial set of parameters, and then use the more accurate three-part approach in a tighter parameter space to finalise the optimisation.

References

1. Lourenço, P., Véstias, Fantoni, A., Vieira, M.: Multimode interference reflectors and output tuning using neural networks. In: Proceedings SPIE 12880, Physics and Simulation of Optoelectronic Devices XXXII, 128800K (2024)
2. Albeladi, F.T., et al.: Multi-mode Interference Reflector based InAs-QD Laser. In: 2022 28th International Semiconductor Laser Conference (ISLC), pp. 1–2, Matsue, Japan (2022). https://doi.org/10.23919/ISLC52947.2022.9943389
3. Kleijn, E., Smit, M.K., Leijtens, X.J.M.: Multimode interference reflectors: a new class of components for photonic integrated circuits. J. Lightwave Technol. **31**(18), 3055–3063 (2013). https://doi.org/10.1109/JLT.2013.2278187
4. Rumpf, R.: Improved formulation of scattering matrices for semi-analytical methods that is consistent with convention. Prog. Electromagn. Res. B. **35**, 241–261 (2011). https://doi.org/10.2528/PIERB11083107

5. Albeladi, F.T., et al.: Design and characterisation of multi-mode interference reflector lasers for integrated photonics. J. Phys. D. Appl. Phys. **56**(38), 384001 (2023)
6. Rumpf, R. Electromagnetic and Photonic Simulation for the Beginner: Finite-Difference Frequency-Domain in MATLAB 2022
7. Yee, K.: Numerical solution of initial boundary value problems involving maxwell's equations in isotropic media. IEEE Trans. Antennas Propag. **14.3**, 302–307 (1966)
8. Demir, V.: Graphics processor unit (GPU) acceleration of finite-difference frequency-domain (FDFD) method. Progress In Electromagnetics Research M. **23**, 29–51 (2012). https://doi.org/10.2528/PIERM11090909
9. Le, T.-T.: An improved effective index method for planar multimode waveguide design on an silicon-on-insulator (SOI) platform. Opt. Appl. **43**, 271–277 (2013)
10. Ivinskaya, A.: Finite-Difference Frequency-Domain Method in Nanophotonics. In: PhD thesis (2011)

Chapter 22
3D Printed Microlenses on Silicon Photonic Chips for Optical Coupling Optimization with Free Space Beams

L. Zerboni, E. Gubello, A. Martinez, G. Cannetti, A. Melloni, and F. Morichetti

Abstract We report on the design and realization of two-photons polymerization (TPP) 3D-printed microlenses integrated onto silicon photonic chips for the optimization of light coupling between optical waveguides and free-space beams. Experimental results show a 5x magnification of the beam radiated by grating couplers, with a divergence reduction to less than 20 mrad and a 15 dB improvement of coupling efficiency with free space beams with hundreds of microns waist.

22.1 Introduction

Efficient optical coupling is a fundamental challenge in integrated photonics, especially when interfacing high refractive-index contrast optical waveguides with optical fibers or free-space optical beams. mode adaptors for edge coupling [1, 2], like inverse tapers [3, 4], or surface grating couplers can be designed and monolithically integrated on chips to optimize the properties of the optical beams, like beam waist, divergence and phase front profile. However, due to the constraints of planar structures, the available degrees of freedom for beam shaping are quite limited.

In this work, we report on the design and realization of microlenses that are directly 3d-printed onto silicon photonic chips in order to improve the coupling efficiency between grating couplers and free-space beams. these microlenses are fabricated using high-resolution two-photon polymerization (TPP) [2, 4, 5], enabling precise control over their geometry to optimize the optical beam shape and minimize coupling losses. A novel lens design is proposed that accommodates the 12.5° tilted radiation of the grating coupler, provides a 5X magnification of the beam, while reducing the

L. Zerboni (✉) · E. Gubello · A. Martinez · A. Melloni · F. Morichetti
DEIB, Politecnico di Milano, Milan, Italy
e-mail: lorenzo.zerboni@polimi.it

G. Cannetti
POLIFAB, Via Giuseppe Colombo, Milan, Italy

P. Smowton and T. Krauss (eds.), *Proceedings of the 26th European Conference on Integrated Optics*, Springer Proceedings in Physics 440,
https://doi.org/10.1007/978-3-032-29673-3_22

divergence to less than 20 MRAD. examples of 3d printing of individual lenses, as well as 1d and 2d arrays of microlenses, are reported. This technique offers a scalable, cost-effective solution for interfacing multi-aperture photonic integrated circuits to free-space and multicore/few-mode fibers, with promising applications in optical communication, sensing, and quantum photonics.

22.2 Lens Design and Fabrication

The photonic chip used in this work was fabricated on a commercial 220 silicon photonics platform. The grating coupler is a standard PDK of the foundry and is optimized for optical coupling with single mode optical fibers. Figure 22.1a_1 shows the finite-difference time-domain (FDTD) simulation of the near field of the grating, which emits the optical beam field at a tilt angle of 12.5°, with a divergence of 5° (elevation) and 9° (azimuthal). Figure 22.1a_2 shows a top-view scanning electron microscopy (SEM) picture of one of the fabricated gratings.

To reduce the divergence of the radiated beam, polymer micro lenses were directly 3D-printed on top of the grating couplers by using TPP technology [4]. The geometry and SEM picture of the realized microlens is shown in Fig. 22.1 b_1 and b_2, respectively. The design parameters of the lens have been optimized by using Fresnel Propagation Method (OpticStudio, Physical Optics Propagator). The lens axis is tilted by 12.5° to be aligned to the tilt of the grating coupler beam and the curvature follows

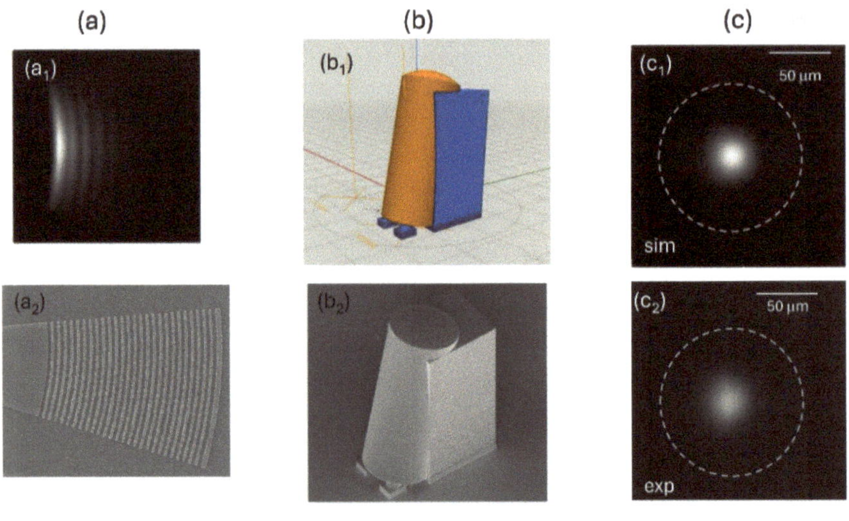

Fig. 22.1 FDTD simulation of the near field of the silicon grating coupler $\mathbf{a}_1$, and top view SEM picture of one of the fabricated gratings $\mathbf{a}_2$. 3D model of the printed microlens $\mathbf{b}_1$ and SEM picture of one of the 3D-printed fabricated lenses. Simulated $\mathbf{c}_1$ and measured $\mathbf{c}_2$ optical beam on the top of the lens. The dashed circle represents the geometrical dimension of the lens (125 µm diameter)

the spherical phase front profile of the beam at the top surface of the lens, while a gap of 30 μm has been kept between the lens and the grating coupler. The lens has a total length of 320 μm, a diameter of 125 μm, and a radius of curvature of 115 μm. The printing process with IP-S resin required about 25 minutes. The numerical simulation and the near-IR camera picture of the beam emerging from the top surface of the lens is shown in Fig. 22.1c_1 and c_2, respectively.

22.3 Lens Performance Assessment

In order to test the optical performance of the fabricated lens, we used the setup shown in Fig. 22.2a. The device under test is a single-mode silicon straight waveguide with input and output grating couplers. On top of the output grating coupler a microlens is printed. Single mode optical fibers (SMF 28) are used to couple the light (at a wavelength of 1550 Nm) into the waveguide.

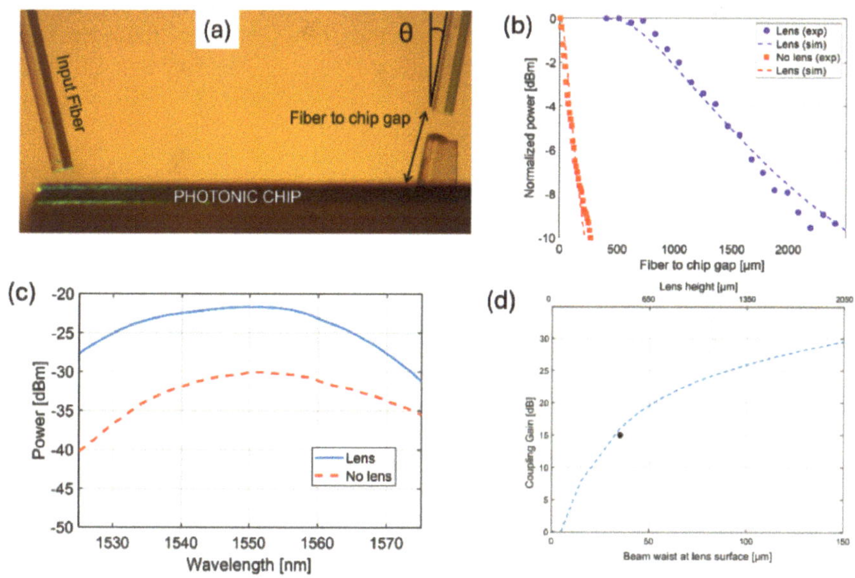

Fig. 22.2 **a** Experimental setup is used to test the optical performance of the lens, where light is coupled into a silicon waveguide using single-mode optical fibers (SMF 28). **b** Simulated (dashed lines) and experimental (markers) power collected by the output fiber as a function of the fiber-to-chip gap distance, with (blue circles) and without (red squares) the integrated microlens. **c** Optical spectrum measured before (dashed red) and after (solid blue) lens printing. **d** Simulated (dashed line) and experimental (black circle) gain provided by the lens along with the required lens height versus the beam waist at the lens surface (free-space beam was generated with a single mode optical fiber placed at a suitable distance from the lens surface)

The first test was conducted to investigate the reduction of the beam divergence provided by microlens. Figure 22.2b shows the normalized power collected by the output fiber versus the gap distance between the fiber tip and the surface of the chip with (blue circles) and without (red squares) a lens integrated onto the output grating coupler (normalization is done with respect to the peak power). Experimental data is in good agreement with numerical simulations (dashed lines). From the slope of the curves the beam divergence can be evaluated. The Rayleigh distance measured without the lens is about 50 μm, while with the lens it increases up to 1.25 mm. The corresponding divergence angle calculated by applying the Gaussian beam approximation is 100 mrad and 20 mrad (half angle) in the case of non-lensed and lensed grating coupler respectively. The reduction of the divergence beam in Fig. 22.2b is related to the beam dimension at the lens output, by $\theta = \lambda/\pi w_0$, where θ represents the beam divergence (half angle), λ is the wavelength, and w_0 the beam waist on the lens surface.

As shown in Fig. 22.2c the introduction of the lens does not significantly alter the optical spectrums of the grating coupler, which maintain a 3 dB bandwidth of 35 nm around 1550 nm. Figure 22.2d shows the coupling efficiency gain given by the lens when coupled with a free space beam generated with a single mode optical fiber placed at a suitable distance from the lens surface, as a function of the beam waist radiated by the lens (the black dot represents a 15 dB experimental gain measured with the fabricated lens).

22.4 Conclusions

In this work, we demonstrated the design, fabrication, and optical characterization of 3D-printed polymer microlenses integrated onto grating couplers in a silicon photonics platform. The microlenses were designed to match the emitted beam profile of the grating couplers, reducing the divergence and enhancing the coupling efficiency. Experimental results show a reduction in the beam divergence from 100 mrad to 20 mrad, in good agreement with numerical simulations, and no significant change in the spectral response (3 dB bandwidth of 35 nm around 1550 nm).

These lenses can be integrated into photonic arrays (see Fig. 22.3) as well as printed in front of edge couplers, or on other optical interfaces. Moreover, they could enable efficient coupling with fiber arrays, few-mode and multi-core fibers, as well as chip-to-chip optical communication expanding the potential applications of integrated photonics. These findings highlight the potential of 3D-printed micro-optics to improve optical interfaces in silicon photonics, enabling more efficient fiber-to-chip coupling and expanding the application possibilities for integrated photonic devices. Future work will also focus on optimizing the lens design for further improvements in performance and exploring different polymer materials for enhanced optical properties.

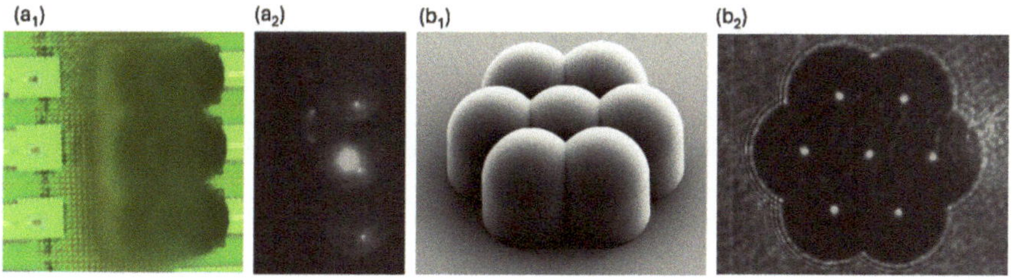

Fig. 22.3 Example of microlenses arranged in linear **a** and circular **b** arrays. $\mathbf{a_1}$ Visible light image focused on the chip surface, showing the printed microlenses. $\mathbf{a_2}$ Infrared image focused on the top surface of the lenses, with thermal light illuminating the lens surface and the central grating emitting light. $\mathbf{b_1}$ SEM image of microlenses printed on a glass wafer in a circular configuration. $\mathbf{b_2}$ Collimated light observed on the glass wafer surface when the lenses were illuminated with a plane wave

Acknoledgments This work was carried out at Polifab, the nanofabrication facility of Politecnico di Milano, and was supported by the Italian National Recovery and Resilience Plan (NRRP) of NextGenerationEU, partnership on "Telecommunications of the Future" (Program "RESTART", Structural Project "Rigoletto," and Focused Project "HePIC") under Grant PE00000001.

References

1. Snyder, B., et al.: Broadband, polarization-insensitive lensed edge couplers for silicon photonics. In: 2018 IEEE 68th Electronic Components and Technology Conference (ECTC), pp. 841–847. San Diego, CA, USA (2018)
2. Ranno, L., Sia, J.X.B., Popescu, C., et al.: Highly efficient fiber to Si waveguide free-form coupler for foundry-scale silicon photonics. Photon. Res. **12**, 1055–1066 (2024)
3. Van Asch, J., et al.: A methodical approach to design adiabatic waveguide couplers for heterogeneous integrated photonics. Journal of Physics: Photonics. **6**(4), 045013 (2024)
4. Blaicher, M., Billah, M.R., Kemal, J., et al.: Hybrid multi-chip assembly of optical communication engines by in situ 3D nano-lithography. Light Sci. Appl. **9**, 71 (2020)
5. Xu, Y., Maier, P., Trappen, M., et al.: 3D-printed facet-attached microlenses for advanced photonic system assembly. Light: Advanced Manufacturing. **4**, 3 (2023)

Chapter 23
An Integrated Source of Pure Single Mode Squeezing

Edward C. R. Deacon, Alexandre Maïnos, Jacob F. F. Bulmer, Molly A. Thomas, Patrick Yard, and Anthony Laing

Abstract We present an integrated source of pure single-mode squeezing on a silicon-on-insulator photonic chip. The source consists of microring resonators that are linearly uncoupled but nonlinearly coupled with the design enabling dual-pump spontaneous-four wave mixing (DP-SFWM) to generate single-mode squeezing while suppressing parasitic noise from single-pump (SP) SFWM. We measure a spectral purity in excess of 97% and significant suppression of SP-SFWM in the device verified with photon counting measurements.

23.1 Introduction

Presented here is an integrated source designed for the generation of single mode squeezing using dual pump spontaneous four-wave mixing (DP-SFWM) that is based on the principle of linearly uncoupled and nonlinearly coupled (LUNC) ring resonators as introduced in [1]. Using ring resonators in a LUNC configuration has several benefits including the suppression of parasitic single pump (SP) SFWM processes which add thermal noise to the desired single-mode squeezed vacuum (SMSV) state; independent control of each rings resonant modes and their FSRs, allowing compensation for self- and cross-phase modulation arising from intense pump fields and simplifying phase matching by reducing the requirement on dispersion engineering; and inherent pump suppression due to the squeezed state and pump fields being confined in different resonators.

There have been several demonstrations of LUNC ring resonators in both the silicon-on-insulator (SOI) and silicon nitride platforms: stimulated FWM has been performed demonstrating the LUNC concept in [2]; DP-SFWM was demonstrated in [3], with the design showing the suppression of SP-SFWM process; and DP-SFWM was demonstrated with the nonlinear coupling region implemented with

E. C. R. Deacon (✉) · A. Maïnos · J. F. F. Bulmer · M. A. Thomas · P. Yard · A. Laing
Quantum Engineering Technology Labs, School of Physics and Department of Electrical, Electronic and Mechanical Engineering, University of Bristol, Bristol BS8 1FD, UK
e-mail: edward.deacon@bristol.ac.uk

P. Smowton and T. Krauss (eds.), *Proceedings of the 26th European Conference on Integrated Optics*, Springer Proceedings in Physics 440,
https://doi.org/10.1007/978-3-032-29673-3_23

a Mach-Zehnder interferometer (MZI), resulting in higher photon pair generation rates, spectral bandwidth and fabrication tolerance compared to directional coupler implementations [4].

To date, no measurements nor optimization of spectral purity have been performed for states of light generated in LUNC ring resonator designs. In the source presented here, we use tuneable couplings on each ring for independent control and optimization of the pump and squeezing resonator linewidths which in conjunction with pulsed pumps results in high spectral purity which is characterized via spontaneous emission tomography (SET) with measurements of the joint spectral intensity (JSI). In addition, we show a high level of extinction of the parasitic SP-SFWM processes via single photon measurements.

23.2 Results

The source is configured such that the top ring, the *pump* ring, and the bottom ring, the *squeezing* ring, are linearly uncoupled via the Mach-Zehnder interferometer (MZI) that adjoins the two rings. The MZI is tuned with a thermal phase shifter to obtain the linear uncoupling. Additionally, the MZIs used to couple the pump and squeezing ring to bus waveguides are tuned to maximize spectral purity; this corresponds to over coupling for the pump ring and under coupling for the squeezing ring to obtain broad and narrow resonance linewidths respectively.

Figure 23.1 depicts a joint spectral intensity (JSI) measured via SET. Performing a Schmidt decomposition on the measured JSI, we extract a spectral purity in excess of 97% showing that spectral correlations in the JSI have largely been eliminated and that the produced state is highly single-moded.

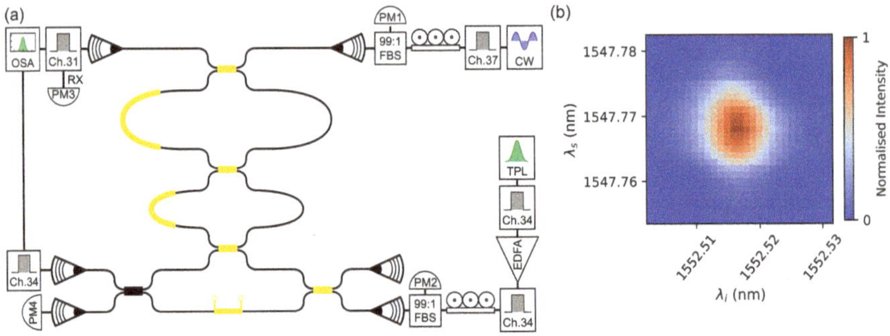

Fig. 23.1 **a** Experimental setup for performing stimulated emission tomography (SET) of the LUNC ring source. OSA: optical spectrum analyzer. CW: Continuous wave laser. TPL: Tuneable pulsed laser. EDFA: Erbium-doped fibre amplifier. **b** Measured Joint spectral intensity of the LUNC ring resonator source measured via SET which is performed with a single pulsed pump laser and a CW seed laser

In addition to the spectral purity of the source, we also characterize its ability to suppress parasitic SP-SFWM that occurs in the DP-SFWM process used to generate SMS. This is performed with photon counting measurements where we use off-chip wavelength division multiplexing (WDM) filters to measure the energy matched channels that are symmetric about the pumps. We simultaneously measure the channels corresponding to the SP- and DP-SFWM and look at coincidences between channels to measure the degree of suppression.

Figure 23.2a depicts the experimental setup consisting of the two pulsed pump lasers and cascaded WDM filters used to separate the energy-matched SFWM channels. In Fig. 23.2b, we observe coincidences between detectors 1 and 2 (D1-D2) corresponding to degenerate photons generated via DP-SFWM to be the strongest. We then consider coincidences between detector channels D1-D3 and D1-D4 which correspond to the parasitic processes. These processes are almost completely suppressed as seen by the ratio of the central to side peaks and compared to the DP-SFWM process, the count rates are over 15 dB lower.

23.3 Conclusion

We have demonstrated a integrated source of single mode squeezing that spectrally pure and low in noise generated from SP-SFWM processes. This is achieved by using a LUNC configuration of ring resonators that are coupled to one another and input/output bus waveguides via MZIs which allow the linear uncoupling and coupling conditions for high spectral purity to be optimized.

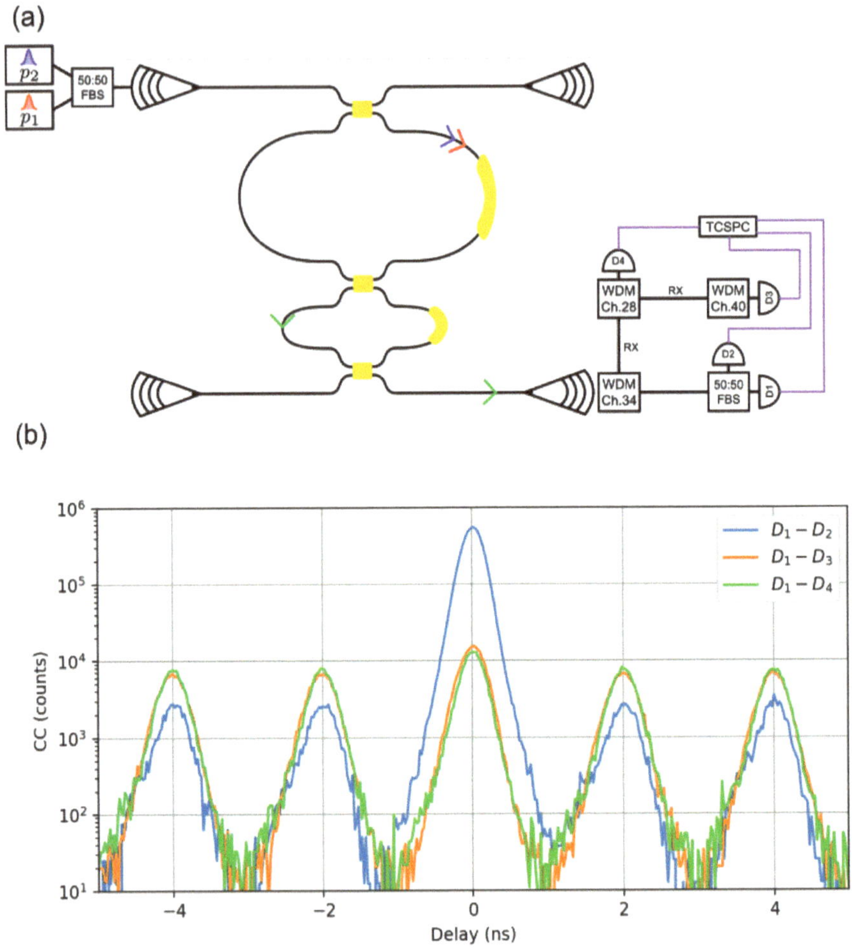

Fig. 23.2 **a** The LUNC ring source design. Two pulsed pumps are injected and coupled to the pump ring via a MZI. A second MZI couples the pump ring to the squeezing ring which is set to identity to linearly uncouple the two rings. Inside this MZI, DP-SFWM occurs generating a SMSV which is resonant with the squeezing ring. The SMSV is then outcoupled of the squeezing ring by a third MZI and filtered and measured off-chip with WDMs and SNSPDs, with photon detections processed using a time tagger

References

1. Menotti, M., et al.: Nonlinear coupling of linearly uncoupled resonators. Phys. Rev. Lett. **122**, 013904 (2019)
2. Tan, K., et al.: Stimulated four-wave mixing in linearly uncoupled resonators. Opt. Lett. **45**, 873–876 (2020)
3. Sabattoli, F.A., et al.: Suppression of parasitic nonlinear processes in spontaneous four-wave mixing with linearly uncoupled resonators. Phys. Rev. Lett. **127**, 033901 (2021)
4. Sabattoli, F.A., et al.: Nonlinear coupling of linearly uncoupled resonators through a Mach–Zehnder interferometer. Appl. Phys. Lett. **121**, 201101 (2022)

Chapter 24
Silicon Photonic Coherent Adder for Communication and Sensing with Time-Multiplexed Control Feedback

Samuele De Gaetano, Monica Crico, Francesco Zanetto, Giorgio Ferrari, Marco Sampietro, Francesco Morichetti, and Andrea Melloni

Abstract Time-multiplexed control of a reconfigurable silicon photonic coherent adder is demonstrated by means of monolithically integrated CMOS electronics. Time-multiplexing is used to interrogate integrated photodetectors and drive thermal actuators with few electrical signals, realizing accurate compensation and locking of mutual phase drifts of the input signals to be combined. The optical circuit functionality is then demonstrated both with a transmission experiment, where a 25 Gbps modulated signal is reconstructed at the chip output, and in a sensing scenario, where the chip is used to assess the phase difference between the input light beams.

24.1 Introduction

The development of large-scale programmable photonic integrated circuits (PICs), like meshes of Mach-Zehnder interferometers (MZI) for optical computing or interconnect applications [1], requires an external control unit to configure the chip functionality, monitor it in real-time and keep it stable against possible drifts and variations, due to aging or environmental effects [2]. Therefore, to ensure adaptive operations and full programmability, these PICs require multiple on-chip light sensors and actuators, that are employed to close the external electronic configuration and stabilization loop [3]. The number of such elements is strictly related to chip complexity and can easily reach the scale of few hundreds for large PICs. Solutions to integrate photonic and electronic functionalities into the same chip have thus been investigated to reduce the number of electrical connections between the PIC and the electronics [4].

S. De Gaetano (✉) · M. Crico · F. Zanetto · M. Sampietro · F. Morichetti · A. Melloni
Department of Electronics, Information and Bioengineering, Politecnico di Milano, Milan, Italy
e-mail: samuele.degaetano@polimi.it

G. Ferrari
Department of Physics, Politecnico di Milano, Milan, Italy

P. Smowton and T. Krauss (eds.), *Proceedings of the 26th European Conference on Integrated Optics*, Springer Proceedings in Physics 440,
https://doi.org/10.1007/978-3-032-29673-3_24

Among programmable PICs, coherent adders based on self-configuring structures are gaining popularity in recent years. They are usually implemented with meshes of MZIs arranged in a diagonal or binary tree configuration. Their credentials have already been validated with demonstrations of optical mode manipulation and free-space optical transmission among others [5, 6]. However, the versatility of these PICs also allows their use in those applications where, in addition to optical transmission and routing, measurements of the input beam characteristics are required (the so-called transmission and sensing approach).

In this work, we demonstrate time-multiplexed control of an MZI-based silicon photonic coherent adder, enabled by monolithically integrated CMOS electronic circuits. Two integrated electronic multiplexers, simultaneously operated for both the readout and actuation tasks, significantly reduce the number of external connections while still maintaining feedback control over the photonic chip. The functionality of the PIC is demonstrated in a transmission and sensing scenario, where a 25 Gbps transmission experiment is carried out together with real-time tracking of the phase difference between the circuit inputs.

24.2 Time-Multiplexed Control of an Adaptive Coherent Adder

The architecture of the PIC, manufactured by a commercial silicon photonics foundry (AMF, Singapore) with a zero-change approach, is illustrated in Fig. 24.1. Each MZI within the binary tree adder is equipped with dedicated germanium photodiodes and thermal phase shifters, all controlled by external electronic circuits. To minimize the number of electrical input/output (I/O) connections required for interfacing the chip with the control electronics, we implemented an on-chip multiplexer (MUX) that sequentially routes the current signals from the integrated sensors towards a single acquisition circuit [7]. After digitization with an analog-to-digital converter, the serialized signals are separated into parallel signal processing chains inside the FPGA that manages the electronic system. The conceptual operating principle of the time-multiplexed readout is shown in Fig. 24.2a. The elaborated information is then serialized again in the FPGA and brought back into the analog domain with a single digital-to-analog converter (DAC), driving a second integrated demultiplexer (DEMUX) that sequentially connects each MZI actuator. A critical aspect of this approach is maintaining the stability of the MZI working point when the DEMUX switches from one actuator to the next one. To achieve this, we have integrated an electronic memory on the chip, which includes a sample-and-hold (S&H) circuit connected to the current drivers of the thermal actuators [7]. The S&H circuit exploits DEMUX switches together with a memory capacitor bank to store the heater voltage information and keep it stable during the DEMUX operation. The time-multiplexed control scheme for the actuators is illustrated in Fig. 24.2b. During the sampling phase, commands are applied to the actuator drivers and memorized by the S&H. In

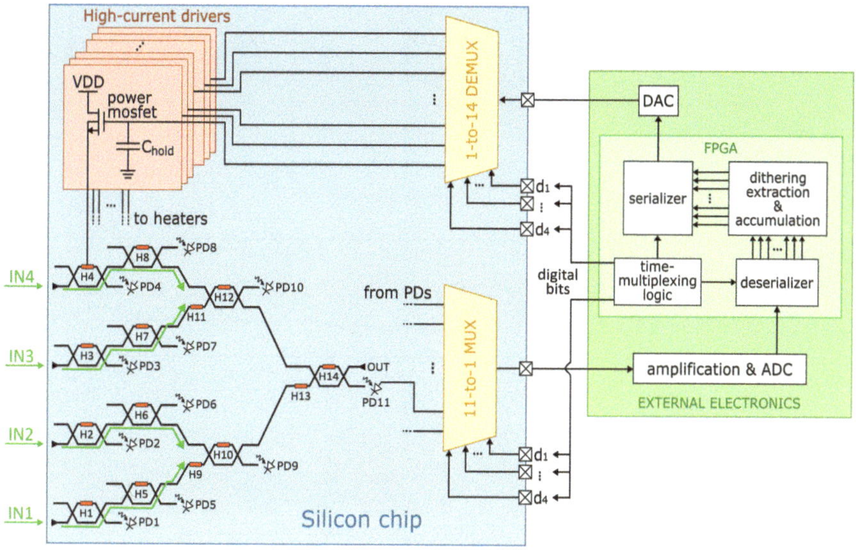

Fig. 24.1 Schematic view of the silicon chip, integrating a photonic coherent adder and two electronic multiplexers enabling sequential control of the optical functionality

the hold phase, the stored charge in the capacitors maintains the actuator commands. Notice that, thanks to the on-chip drivers, a low power external DAC is enough to operate the circuit.

The fully time-multiplexed control, only theoretically anticipated in [7], was exploited to operate the MZI mesh as an adaptive coherent adder. The experimental setup consists of an optical fiber array for signal injection, a custom printed circuit board (PCB) serving as an interface to the external control electronics, and a single fiber for output signal retrieval (Fig. 24.3a). A 1550 nm laser source was equally divided by an external 1:4 fiber splitter and injected into the photonic circuit through a fiber array. Two cascaded MZI stages (first and second stages) act as high-speed optical switches to route the input signals towards the following 4 × 1 MZI mesh (third and fourth stages), functioning as a coherent adder for signal summation. Output power was coupled from the circuit via a grating coupler. To achieve autonomous MZI calibration without prior system characterization, 2 kHz dithering modulation scheme (as demonstrated in [5]) has been employed. This calibration process operates with MUX/DEMUX components cycled at 156 kHz, enabling sequential interrogation of individual channels at around 10 kHz.

Figure 24.3b shows the convergence curve of the photocurrent of the monitor PDs during the automatic configuration of the MZI mesh. Initially, the working points of all MZIs were set by applying random voltages to the integrated heaters. The activation of the feedback loops resulted in the minimization of the PD photocurrents by precisely adjusting phase shifter voltages and maximization of the optical power at

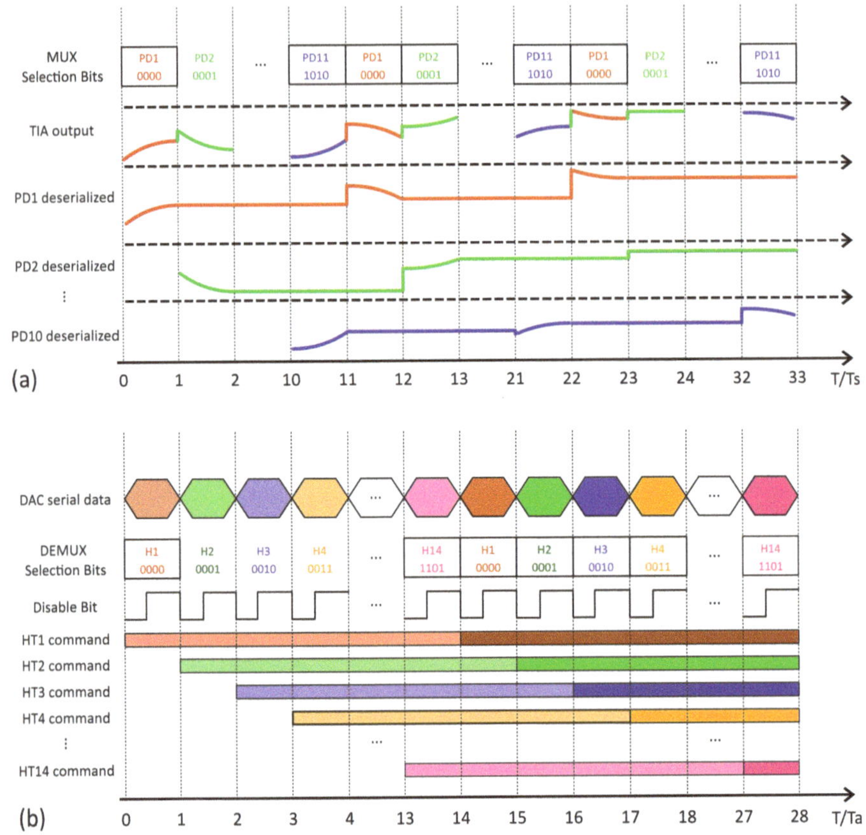

Fig. 24.2 Time-multiplexed algorithm for sequential control of **a** multiple sensors and **b** phase actuators, using a single analog input and output

output port (OUT). Each MZI achieves stable operation within tens of milliseconds, enabling the full configuration of the entire PIC in under 50 ms.

24.3 25 Gbps Transmission Experiment

The experimental validation of the automated time-multiplexed control system has been carried out in a high-speed data transmission scenario. To perform the experiment, the laser beam has been modulated before the fiber splitter and the resulting four signals simultaneously injected into the chip. After reconstruction by the coherent adder, the output signal has been monitored with an optical oscilloscope. Figure 24.4 illustrates comparative eye-diagram analyses of a 25 Gbps intensity-modulated signal from the MZI mesh under different operational conditions.

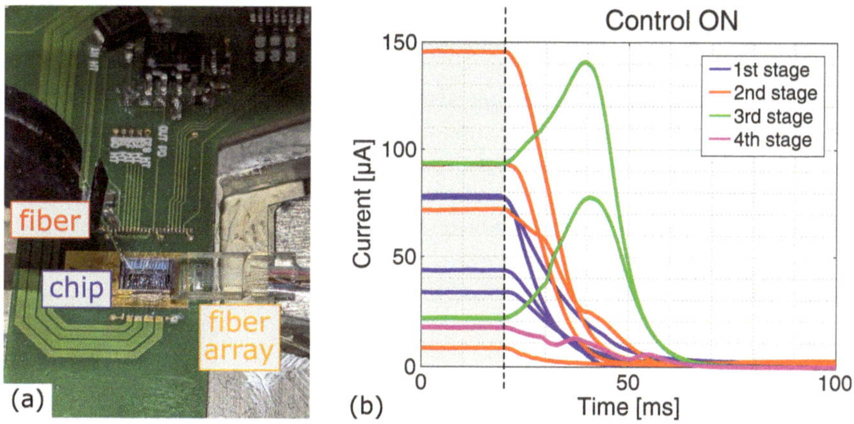

Fig. 24.3 **a** Electro-optical setup consisting of an input fiber array, integrated photonic chip, electronics interfaces and output fiber. **b** Convergence curves of the feedback control system, showing the minimization process of the photodiode signals across four stages

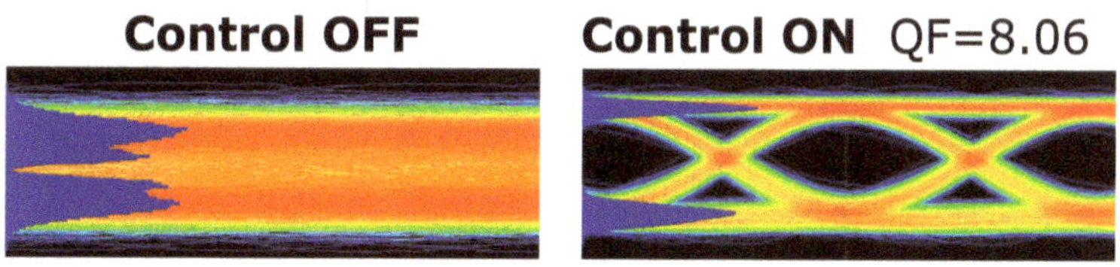

Fig. 24.4 Eye diagrams of a 25 Gbps intensity-modulated signal from the coherent adder output when the active phase stabilization is disabled (left) and enabled (right)

Without active electronic stabilization, environmental perturbations such as thermal drifts, acoustic interference, and mechanical vibrations in the input fiber array introduce time-dependent phase drifts among the input beams of the coherent adder. Therefore, the four signals are randomly combined by the circuit and the transmitted information is completely lost, resulting in a completely closed eye diagram. On the other hand, the activation of the closed-loop control system restores optimal beam coherence, producing a well-defined open eye diagram with a measured quality factor of Q = 8.06. Successful compensation of the random phase fluctuations is achieved within the 50 ms response of the control loop, enough to mitigate the slow drifts that usually affect optical fiber components.

24.4 Input Phase Sensing

As previously discussed, the coherent adder can be used both for transmission and sensing purposes. In particular, the time-dependent variations in the relative phase of the input beams, caused by external factors, can be evaluated by analyzing the operating point of the MZIs. Indeed, to achieve coherent summation of the input optical powers, the phase shifts introduced by the thermo-optic actuators must satisfy the following conditions:

$$\Delta\theta = 2\tan^{-1}\left|\frac{x_1}{x_2}\right|, \theta \in [0, \pi]\ \Delta\phi = (\angle x_1 - \angle x_2) + 2N\pi \tag{24.1}$$

where x_1 and x_2 are the optical fields at the input of an MZI (Fig. 24.5a).

Since our control strategy is based on active feedback, it is possible to continuously monitor the electrical power supplied to the input thermo-optic actuators to retrieve in real-time the phase mismatch $\Delta\phi$ between the optical input beams of the 4×1 binary tree (Fig. 24.5b). The computed phase mismatch is relative to a fixed reference at a given time instant t_0.

This idea has been experimentally verified. Figure 24.5c shows the evolution of the phase difference between the inputs of the coherent adder (third stage of the PIC), obtained by analyzing the heater command generated by the feedback loop. The measurement confirms the presence of phase fluctuations due to environmental effects, that can be as high as 0.4π on a time span of few minutes. The continuous control action, performed by the feedback electronics, thus allows to both estimate the magnitude of this phenomenon and to counteract it in real-time.

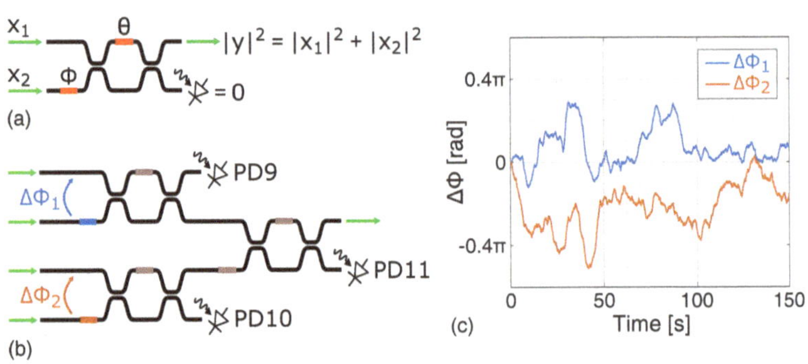

Fig. 24.5 **a** MZI used in a coherent adder scheme. **b** 4×1 binary tree of MZIs used to evaluate the input phase variations, thanks to the active feedback loop that constantly routes light to the output of the circuit. **c**) Long-term experimental measurement of the input phase drifts, showing time-dependent variations in the relative phase shifts ($\Delta\phi_1$, $\Delta\phi_2$)

24.5 Conclusions

We presented the electronic control of a programmable silicon PIC where, for the first time, on-chip time-multiplexing is used both for interrogating integrated monitor PDs and driving volatile thermal actuators. The proposed methodology requires no modifications to commercial silicon platforms and minimizes the number of electrical I/Os, thus enabling scalability to larger photonic chips. The effectiveness of the approach is demonstrated by accurate and stable configuration of an integrated MZI-based coherent adder, achieved by real-time compensation of phase drifts in the input signals. The feedback-controlled chip has been used for both communication and sensing purposes, with the transmission of a 25 Gbps signal together with the estimation of the phase difference between the input beams. The latter information could be useful, for instance, in free-space applications to estimate the angle of arrival of the beam impinging on the coherent adder inputs without the need of complex algorithms.

Acknowledgments This work was supported by the Italian National Recovery and Resilience Plan (NRRP) of NextGenerationEU, partnership on "Telecommunications of the Future" (Program "RESTART", Structural Project "Rigoletto," and Focused Project "HePIC") under Grant PE00000001. The authors thank Polifab, the nanofabrication facility of Politecnico di Milano, for chip assembly.

References

1. Bogaerts, W., et al.: Programmable photonic circuits. Nature. **586**(782), 207–216 (2020)
2. Milanizadeh, M., et al.: Control and calibration recipes for photonic integrated circuits. IEEE J Sel Top Quantum Electron. **26**(5), 1–10 (2020)
3. Ribeiro, A., et al.: Column-row addressing of thermo-optic phase shifters for controlling large silicon photonic circuits. IEEE J Sel Top Quantum Electron. **26**(5), 1–8 (2020)
4. Stojanovic, V., et al.: Monolithic silicon-photonic platforms in state-of-the-art CMOS SOI processes. Opt. Express. **26**(10), 13106–13121 (2018)
5. SeyedinNavadeh, S., et al.: Determining the optimal communication channels of arbitrary optical systems using integrated photonic processors. Nat. Photonics. **18**(2), 149–155 (2024)
6. Martinez, A.I., et al.: Self-adaptive integrated photonic receiver for turbulence compensation in free space optical links. Sci. Rep. **14**(1), 20178 (2024)
7. Zanetto, F., et al.: Time-multiplexed control of programmable silicon photonic circuits enabled by monolithic CMOS electronics. Laser Photonics Rev. **17**(11), 2300124 (2023)

Chapter 25
Agile All-Optical Wavelength Conversion with an on-Chip InP Single-Gain Multi-wavelength Laser

Pablo Marin-Palomo and Martin Virte

Abstract We demonstrate all-optical wavelength conversion over a 1.3 THz range for signals of up to 10 GBd using cross-gain modulation (XGM) in an integrated multi-wavelength laser (MWL). Our approach enables agile control of the wavelength shift without requiring additional probe lasers while achieving high conversion efficiency. The MWL, designed with a dual-cavity structure and a single gain section, supports dynamic wavelength switching at the nanosecond time scale. These features make our solution highly suitable for WDM networks.

25.1 Introduction

Wavelength conversion is key for wavelength division multiplexing (WDM) networks [1], enabling dynamic wavelength allocation, efficient spectral utilization and improved network flexibility. In current optical networks, routing and switching often involve optoelectronic conversion, which introduces significant power consumption and latency due to the need for optoelectronic conversions [2]. To address this bottleneck, all-optical wavelength conversion techniques have been extensively explored, mostly relying on nonlinear effects in semiconductor lasers or amplifiers, such as four-wave mixing and difference-frequency generation [3]. While these techniques allow for broad tunability and transparent data conversion, they require additional probe lasers [4], strict phase matching conditions [1], and suffer from low conversion efficiency, making them impractical for scalable high-speed WDM system. Cross-gain modulation (XGM) in semiconductor optical amplifiers (SOA) has also been considered, offering a simple, phase-matching-free configuration with potential signal gain during the conversion process [5]. However, XGM still

P. Marin-Palomo (✉) · M. Virte
Brussels Photonics Team (B-PHOT), Vrije Universiteit Brussel, Brussel, Belgium
e-mail: pablo.marin-palomo@vub.be

M. Virte
e-mail: martin.virte@vub.be

P. Smowton and T. Krauss (eds.), *Proceedings of the 26th European Conference on Integrated Optics*, Springer Proceedings in Physics 440,
https://doi.org/10.1007/978-3-032-29673-3_25

requires dedicated pump lasers for each wavelength, increasing system complexity and power consumption. In this work, we propose an innovative approach that leverages a custom-designed multi-wavelength laser (MWL) with a single-gain section that provides emission at the desired wavelengths, thereby eliminating the need for separated probe signals [6, 7]. Using our MWL, we demonstrate wavelength conversion of an OOK signal at data rates of up to 10 GBd from a wavelength of 1537.5 nm to 1548.2 nm, corresponding to approximately 1.3 THz offset.

Exploiting mode coupling within the single-gain section of our MWL we demonstrate dynamic control over the emission of various Fabry-Pérot (FP) modes, spaced by approximately 30 GHz. With a feedback cavity, monolithically integrated with the MWL, we achieve precise selectin of the converted wavelength by adjusting the feedback phase, potentially allowing agile wavelength conversion at the nanosecond time scales [6]. Additionally, for specific feedback phase values, we observe a broadcasting effect, where the input signal is simultaneously wavelength-converted across multiple modes.

Given the scalable nature of our MWL approach, we believe that by optimizing the laser cavity design, it is possible to extend the system to support a larger number of modes and higher symbol rates over even larger frequency offsets, constrained only by the SOA gain bandwidth, which exceeds 5 THz in InP-based platforms. This lays the foundation for highly integrated and dynamically reconfigurable all-optical wavelength conversion solutions for WDM networks.

25.2 On-Chip Multi-wavelength Laser and Feedback Control

Our multi-wavelength laser (MWL) is fabricated on an InP generic photonic platform, integrating a single gain medium within a dual-cavity laser structure, see Fig. 25.1a for the laser schematic and Fig. 25.1b for the image of the photonic integrated circuit (PIC). The cavity is defined on one side by two distributed Bragg reflectors (DBRs) and on the other side by a broadband reflector based on a multimode interference reflector (MIR), as illustrated in Fig. 25.1a [5]. The two DBRs are spectrally detuned by approximately 10.5 nm (13 THz) and are placed in series to enable selective multi-wavelength emission.

Under a drive current of 50 mA, which is approximately twice the laser threshold, the MWL can emit at multiple lasing modes centered at $\lambda_1 = 1547.94$ nm, $\lambda_2 = 1548.22$ nm, and $\lambda_3 = 1545.50$ nm, see Fig. 25.1c. These modes correspond to the cavity formed by DBR2 and are the dominant modes of this laser due to them featuring a higher gain. At lower wavelengths, the Fabry-Pérot modes around 1537.5 nm which are associated with the cavity defined by DBR1 are the suppressed modes.

As already demonstrated [6], switching between emission at λ_1, λ_2 and λ_3 can be achieved through optical feedback by adjusting the feedback phase under the short

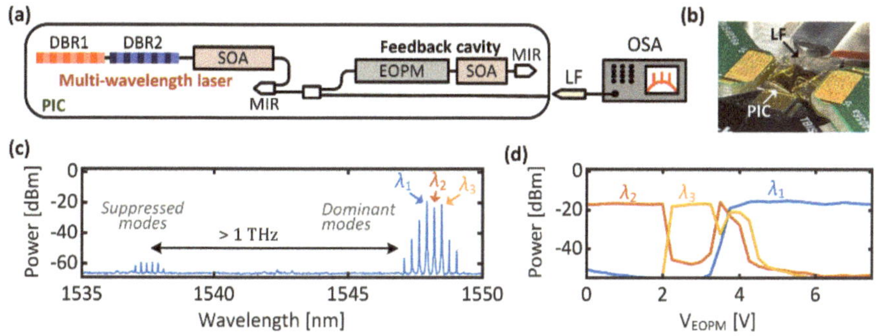

Fig. 25.1 Feedback-based emission control from an on-chip multi-wavelength laser (MWL). **a** Scheme describing the components of our MWL. DBR: Distributed Bragg reflector; SOA: Semiconductor optical amplifier; MIR: Multimode interference reflector; EOPM: Electro-optic phase modulator; LF: Lensed fiber; OSA: Optical spectrum analyzer. **b** Image of the PIC containing several MWL structures. **c** Optical spectrum of MWL output for $I_{FB} = 9$ mA and $V_{EOPM} = 3.75$ V. **d**) MWL output power as a function of the FB voltage V_{EOPM} for $I_{FB} = 9$ mA, $\lambda_1 = 1547.94$ nm, $\lambda_2 = 1548.21$ nm, and $\lambda_3 = 1548.49$ nm

feedback cavity regime. For this reason, a feedback cavity consisting of an electro-optic phase modulator (EOPM, $V_\pi = 9$ V) and an SOA is monolithically integrated with the MWL. By applying voltage V_{EOPM} to the EOPM we are able to adjust the feedback phase, while the feedback strength κ_{FB} is adjusted by applying current I_{FB} to the SOA. Figure 25.1d shows the optical power associated with the various modes emitted by the MWL as we change the feedback phase for $I_{FB} = 9$ mA.

We have shown that switching between the various modes can be achieved at the nanosecond time scale [6]. In addition, for various voltage values, simultaneous emission of different laser modes is observed.

25.3 Data Wavelength Conversion over 1.3 THz in an MWL

Data wavelength conversion is achieved via cross-gain modulation (XGM) by injecting an optical signal around one of the suppressed modes of our MWL. Due to the coupling between the modes of our MWL, gain modulation induced by the injected optical signal is transferred to the un-injected modes which, in our case, are separated by more than 1 THz. We chose to inject the optical signal around one of the suppressed modes to achieve higher wavelength conversion efficiency. In our experiments, we inject the signal around the (FP) mode emitting at 1537.5 nm, although this approach is also effective when injecting around other suppressed FP modes.

We consider an on-off keying (OOK) signal which is generated using an arbitrary waveform generator (AWG) and encoded onto an optical carrier at 1537.5 nm via a Mach-Zehnder modulator (MZM), see Fig. 25.2a. The polarization and power of the optical signal is adjusted using a polarization controller (PC) and a variable optical

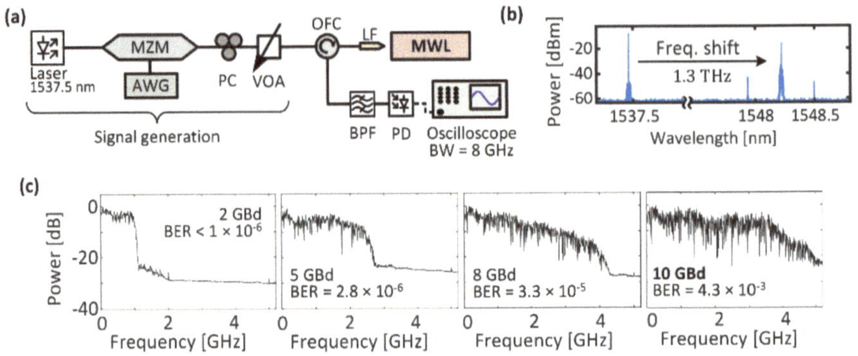

Fig. 25.2 Data wavelength conversion in an MWL. **a** Setup for data wavelength conversion. MZM: Mach-Zender modulator; AWG: Arbitrary waveform generator; PC: polarization controller; VOA: Variable optical attenuator; OFC: Optical fiber circulator. LF: Lensed fiber. BPF: Bandpass filter. PD: Photodetector. **b** Exemplary optical spectrum of the light emitted by the MWL when a 5GBd OOK signal is optically injected around 1537.5 nm. The wavelength-converted signal is observed at $\lambda_2 = 1548.21$ nm. **c** Power spectrum of the OOK signal recorded by the oscilloscope for different symbol rates

attenuator (VOA). A lensed fiber (LF) is used to couple the light into and out of our integrated MWL. Using an optical fiber circulator (OFC) we direct the emission of our MWL towards a photodetector followed by an oscilloscope. An optical bandpass filter (BPF) is used to suppress the regenerated signal still present at the wavelength of 1537.5 nm of the injected signal.

Figure 25.2c illustrates the optical spectrum of the MWL output when an OOK signal at 5 GBd is injected around 1537.5 nm. For data rates beyond 5 GBd, noticeable bandwidth limitations appear, however we demonstrate wavelength conversion up to 10 GBd. We believe that optimizing laser parameters such as drive current and modal gains and injection parameters like detuning and strength could improve signal quality. To investigate these effects, we perform numerical simulations based on laser rate equations to analyze the impact of both laser and optical injection parameters.

25.4 Dynamic all-Optical Data Wavelength Conversion

To achieve data wavelength conversion to any of the wavelengths emitted by our MWL, we optically inject the OOK signal at a wavelength around one of the suppressed modes, see Fig. 25.1b. Mode coupling within the common gain medium enables spectral multiplication of the injected data [5] to the different dominant modes resulting in a conversion over a 1.3 THz offset. Note that the mode competition can be adjusted through both laser and injection parameters allowing for a wide range of possibilities with regards to which FP modes will emit.

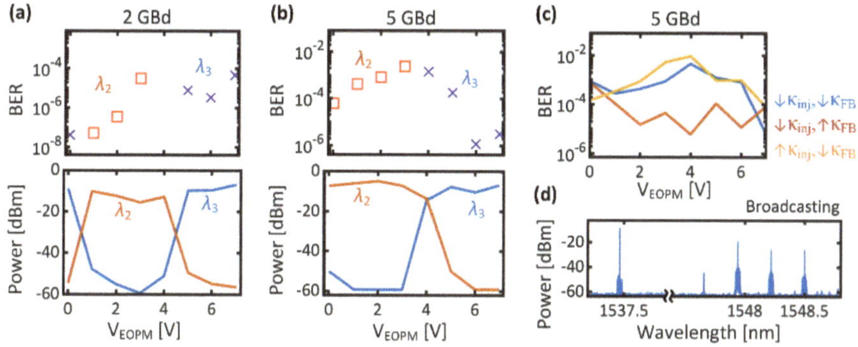

Fig. 25.3 THz-range dynamic all-optical data wavelength conversion. **a** and **b** Top: Measured BER values for each of the 2 channels when varying the feedback phase through V_{EOPM}. **a** and **b** Bottom: Optical power of the corresponding channels. **c** Measured BER values for different configurations of feedback strength κ_{FB}, and optical injection strength κ_{inj}. **d** Broadcasting of the optical injected signal into three channels

To select the emitting mode, we rely on adjusting the feedback phase from a monolithically integrated feedback cavity. By applying a small current to the SOA and a voltage between 0 and 7 V to the EOPM, we can adjust the feedback phase and thus achieve switching between the channels at λ_2, and λ_3, potentially at the nanosecond time scale [6]. The signal quality of the wavelength-converted data is evaluated through BER measurements. Figure 25.3a and b show the BER results for symbol rates of 2 GBd and 5 GBd, respectively, with color and shape-coded data for each channel. As the voltage varies, we observe switching between the two channels, as indicated by the modal power evolutions shown in the lower plots of Fig. 25.3a and b. Additionally, we experimentally analyze the impact of the injection strength and feedback strength, as depicted in Fig. 25.3c. Preliminary results suggest that a higher feedback strength and lower injection strength may improve signal quality, although a more systematic investigation is required to fully characterize this effect.

Furthermore, our approach enables simultaneous broadcasting of the data signal across multiple wavelengths, see Fig. 25.3d.

25.5 Discussion

We have demonstrated all-optical wavelength conversion over a 1.3 THz range for signals up to 10 GBd using cross-gain modulation (XGM) in an integrated multi-wavelength laser (MWL). Our approach enables agile wavelength shifting without requiring additional probe lasers while achieving a high conversion efficiency.

The MWL, designed with a dual-cavity structure and a single gain section, supports dynamic wavelength switching on the nanosecond time scale, making it highly suitable for WDM networks and optical switching applications. Optimizing injection and laser parameters, such as injection detuning and driving current, may allow for higher symbol rates and increase power conversion efficiency. As our approach is based on optical injection, extending the bandwidth to tens of GHz may be feasible with further optimization [8]. While our work focuses on intensity-modulated OOK signals, the feasibility of supporting phase-sensitive modulation formats via cross-phase modulation (XPM) [9] remains an open question.

References

1. Yoo, S.J.B.: Wavelength conversion technologies for WDM network applications. J. Light. Technol. **14, 6**, 955 (1996)
2. Eldada, L.: Advances in ROADM technologies and subsystems. Proc. SPIE. **5970**, 597022 (2005)
3. Wiesenfeld, J.M., et al.: Wavelength conversion at 10 Gb/s using a semiconductor optical amplifier. IEEE Photon. Tech. Lett. **5**, 11 (1993)
4. Hwang, S.K., et al.: All-optical frequency conversion using nonlinear dynamics of semiconductor lasers. Opt. Lett. **34**, 812 (2009)
5. Obermann, K., et al.: Performance analysis of wavelength converters based on cross-gain modulation in semiconductor-optical amplifiers. J. Lightwave Technol. **16**, 1 (1998)
6. Abdollahi, S., et al.: Agile THz-range spectral multiplication of frequency combs using a multi-wavelength laser. Nat. Commun. **15**, 1305 (2024)
7. Marin-Palomo, P., et al.: Data wavelength conversion over THz range using an optical-feedback-controlled InP multi-mode laser. In: Integrated Photonics Platforms III, vol. 13012. SPIE (2024)
8. Liu, Z., Slavík, R.: Optical injection locking: from principle to applications. J. Lightwave Technol. **38**(1), 43–59 (2020)
9. Durhuus, T., et al.: All-optical wavelength conversion by semiconductor optical amplifiers. J. Lightwave Technol. **14**(6), 942–954 (1996)

Chapter 26
Non-linear Effects in Silicon Photonics Microring for Reservoir Computing: Modeling and Experiments

Salvatore Salpietro, Marco Novarese, Cristina Rimoldi, and Mariangela Gioannini

Abstract We investigate how the non-linearities in silicon micro-rings can be exploited for reservoir computing. We present modelling and experiments helpful to predict the performance of the ring pumped by a pump signal that contains the data trace for classification tasks. We analyse how recombination of free carriers via silicon traps determines the carrier lifetime and hence the processing speed and investigate what is the optimal pump level to optimize the decision accuracy.

26.1 Introduction and Motivations

Nonlinear properties of silicon optical micro-ring resonators can be exploited as the non-linear block of a neural network capable of solving classifications tasks using the reservoir computing technique, as demonstrated for the first time in [1]. Silicon micro-rings respond, to a few mW of input bus power, with a variation of the Q factor and a shift of the ring resonant wavelength caused by two-photon absorption (TPA), free-carrier absorption (FCA) and dispersion (FCD), and self-heating (SH). The interplay of these effects rules the choice of the optimal input power and limits the speed of the data processing. To analyse the ring response to an input data set and understand what are the physical limitations to the processing speed and classification accuracy, we have developed a time domain model of the non-linear response of the ring including TPA, FCA, FCD, SH and Shockley Read Hall (SRH) recombination. In parallel, to support our analysis, we have employed the set-up of [2] to reproduce [1] and generate the experimental results that validate the model and support our analysis. As a result we demonstrate that the processing speed is limited by the lifetime of the free-carrier influenced by the SRH recombination rate suggesting that other ring designs with shorter lifetime [2] could be considered to increase the

S. Salpietro · M. Novarese · C. Rimoldi · M. Gioannini (✉)
Dipartimento di Elettronica e Telecomunicazioni, Politecnico di Torino, Torino, Italy
e-mail: mariangela.gioannini@polito.it

P. Smowton and T. Krauss (eds.), *Proceedings of the 26th European Conference on Integrated Optics*, Springer Proceedings in Physics 440,
https://doi.org/10.1007/978-3-032-29673-3_26

processing rate. We also observe an optimum input pump power that maximize the accuracy of the tack; the optimum power level is determined by the fact that too high input can saturate the response of the ring.

26.2 Methods

We used a silicon micro-ring resonator in an add-drop configuration with a quality factor of approximately 13000. We have considered two input datasets: the Iris flower dataset for recognition task and random bits for 1-bit delayed XOR operation [1]. The dataset is encoded in the intensity of the input pump (that excites the silicon non-linearities), while the non-linear response is monitored via sampling, at the output of the MRR, a CW probe signal aligned at the wavelength of another resonance of the MRR. The experimental set-up is summarized in Fig. 26.1. In the experiments, the laser pump is modulated by a Mach-Zehnder Interferometer (MZI) modulator driven by the voltage signal that encodes the input dataset pre-processed by applying a mask. The data of the sampled probe signal are used to train and test the neural network.

The time domain non-linear response of the MRR is modelled by a set of first-order differential equations. The dynamics of the optical electric field circulating in the ring ($E_{c,i}$) at the pump ($i = pump$) or probe ($i = probe$) wavelengths λ_i is given by:

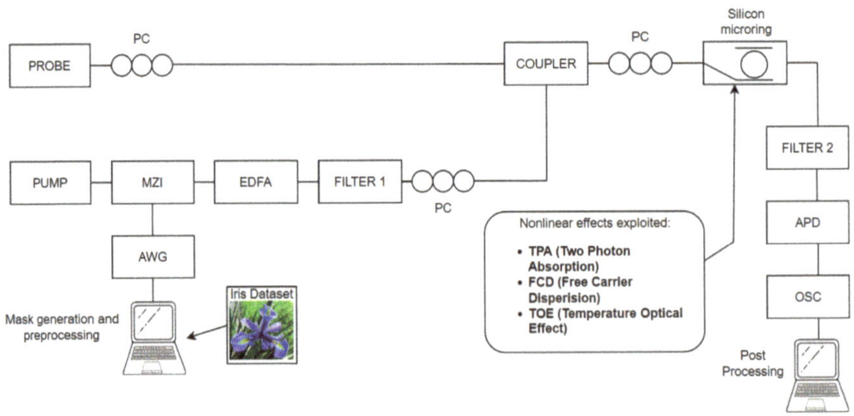

Fig. 26.1 Experimental setup for the pump-probe measurement The configuration includes: Pump and Probe laser sources, MZI modulator and Arbitrary Waveform Generator (AWG), Erbium-Doped Fiber Amplifier (EDFA), various filters, coupler, Polarization Controllers (PC), Avalanche Photodiode (APD) and Oscilloscope (OSC). Control with is for mask generation, preprocessing and post-processing

$$\frac{\partial E_{c,i}}{\partial t} = \left[-\frac{\kappa_i}{t_i^2 \cdot a_i(t)} E_{\text{bus},i} + E_{c,i}(t) \left(\frac{1}{t_i^2 \cdot a_i(t)} - 1 \right) \right] / \tau_g + \left(-i \cdot \frac{2\pi L}{\lambda_i} \cdot \Delta n_{\text{eff},i}(t) \cdot E_{c,i}(t) \right) / \tau_g, \tag{26.1}$$

with κ_i the ring coupling coefficient and $t_i^2 = 1 - \kappa_i^2$; τ_g the group delay per round trip, $\Delta n_{\text{eff,i}}$ the variation of the ring effective refractive index caused by FCD or self-heating, $a_i(t)$ the loss per round-trip caused by TPA, FCA and linear loss. The instantaneous variation of effective refractive index and loss depends on the instantaneous free density of electrons $n_e(t)$ and holes $p_e(t)$ found solving the rate equation for carrier generation and recombination:

$$\frac{dp_e}{dt} = G - \frac{1}{\tau_{p0}} \left(\frac{(p_0 + p_1 + p_e)(p_e - n_e)}{N_f} + \frac{p_e p_1}{p_e + p_0} \right), \tag{26.2}$$

$$\frac{dn_e}{dt} = G - \frac{1}{\tau_{n0}} \left(\frac{(n_0 + n_1 + n_e)(n_e - p_e)}{N_f} + \frac{n_e n_1}{n_e + n_0} \right), \tag{26.3}$$

G is the generation rate due to TPA, while the other second term at the right hand side is the SRH non-radiative recombination rate. Further, $n,\ p_0$ and $n,\ p_1$ are the electron and hole concentrations under equilibrium conditions; N_f is the trap density in silicon, $\tau_{n0,p0}$ are the capture times in the trap state of electrons and holes respectively. The dynamics of temperature variation is modelled via an equivalent multi-node ($j =$ 1, ..., 4) electric circuit whose thermal resistances (R_j) and capacitances (C_j) where derived by 3D thermal simulations previously carried on in Comsol Multyphysics.

$$\frac{d\Delta T_j(t)}{dt} = -\frac{\Delta T_j(t)}{R_j C_j} + \frac{P_{\text{abs},pump}(t) + P_{\text{abs},probe}(t)}{C_i}, \tag{26.4}$$

Here $\Delta T_j(t)$ is the temperature variation at each node ($j = 1...4$) and the silicon waveguide temperature variation is the sum of the temperature variation all-over the nodes. The second term at the RHS is the heat generation corresponding to the absorbed power ($P_{abs,i}$) of the pump and probe signals.

26.3 Results

The model is validated in Fig. 26.2 by comparing the measured and simulated probe traces out of the ring, when the input pump power is a random bit sequence at wavelength corresponding to one cold resonance of the ring and the probe is a weak CW signal at the other nearest resonance. Average pump power is 6dBm, ER=13 dB and bit rate is 10 Mbit/s. The output probe is then sampled to perform the 1-bit delay XOR logic function. Both simulation and experiments show that the output response to one bit depends on the previous history because of the long recovery

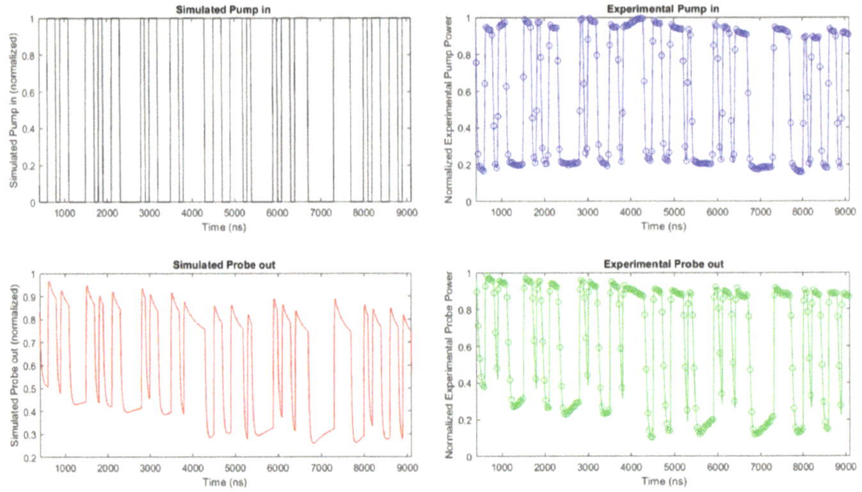

Fig. 26.2 Comparison between simulated (left) and experimental (right) "pump in" and "probe out" signals for the 1-bit delayed XOR experiment. The curves are normalized between 0 and 1

of the absorption and refractive index variation after a pump pulse. The recovery is determined by the SRH recombination and hence by the carrier lifetime. For example, the decrease of the probe power after long sequence of ones is due the recombination of carriers after the circulating optical field is reduced because of the blue shift (sustained by the pump) of the ring resonance. When a zero arrives, the carrier recombination brings the resonance back at the probe wavelength and the output is minimum. For a long sequence of zeros, the slow increase of the probe is due to partial self-heating. The overall decrease of the probe with time is due to the long thermal transient ruled by the longest thermal time constant associated to the low thermal conductivity of silicon dioxide.

In the post-processing phase, the matrix X, which contains the sampled values of the probe signal at the output of the microring (virtual nodes), is reorganized so that each column of the resulting matrix contains the nonlinear transformation of a bit and its preceding bit. Specifically, the first three rows contain the virtual nodes corresponding to the current bit, while the lower three rows contain the virtual nodes corresponding to the previous bit.

This type of structure ensures that each column corresponds to a target XOR result, enabling supervised learning. This allows the computation of the weight matrix W_y using ridge regression. Once trained, it is applied to new samples in the testing phase. On the results, the "Winner Takes All" technique is used to perform the final prediction.

The accuracy obtained by processing the probe signal at the output of the microring resonator is 100%, in both experimental and simulated cases. By applying the

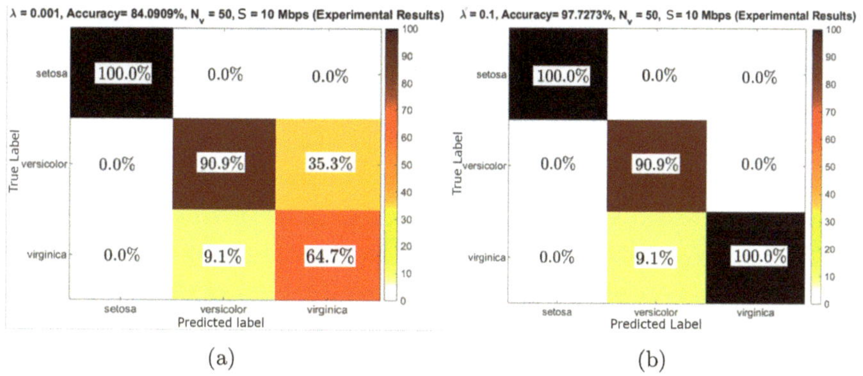

Fig. 26.3 **a** Error map obtained from sampling the input pump signal before the microring. **b** Error map obtained from sampling the output probe signal after the microring

decision algorithms to the modulation voltage the accuracy is about 50% in both simulations and experiments; it increases to 83% out of the MZI for the non-linearity introduced by the modulator.

In the case of the Iris flower dataset, Fig. 26.3 compares the measured classification accuracy with data at 10 Msample/s when the network is trained on the input pump signal (Fig. 26.3a) or output probe (Fig. 26.3b). The non-linearity introduced by the micro-ring is beneficial in increasing the classification accuracy of the virginica species; the classification accuracy for setosa and versicolor are yet high in Fig. 26.3a because the pump signal is already affected by some non-linear distortion introduced by the MZI modulator. If the classification was performed on the modulating voltage, the accuracy would be much lower for all the three species. To investigate quantitatively how much the non-linearity introduced by the ring impacts on the accuracy, we have repeated in simulations with our model the classification of the Iris dataset with data at 30 Msample/s and different average pump powers. The result in Fig. 26.4 shows an optimum power range for the best accuracy. If power is too low, the ring response remains almost linear; however if the power pulse is too high the ring is moved too much out of resonance and the return to the cold resonance for low power levels takes too long. This is further analyzed in Fig. 26.5 where we plot some characteristic parameters as the pump power varies with time, with an average pump power of 15 dBm, a sampling rate of 30 Msps and 50 virtual nodes.

At this high level of average pump power, when the electron density rapidly decreases, the shift of the resonance wavelength due to free carrier dispersion quickly cancels out, while the shift due to the temperature decreases more slowly. So, the stronger heating effect, induced by the high pump power, increases the time required for the system to cool down, keeping the probe resonant wavelength far from the cold probe resonance for a longer period. When the resonance wavelength is significantly shifted away from the cold resonance, the circulating power inside the microring

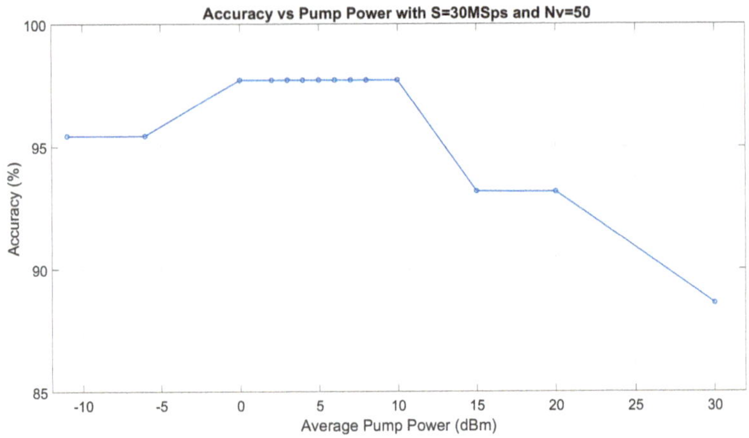

Fig. 26.4 Accuracy as a function of average pump power for a sampling rate of 30 MSps and $N_\upsilon = 50$

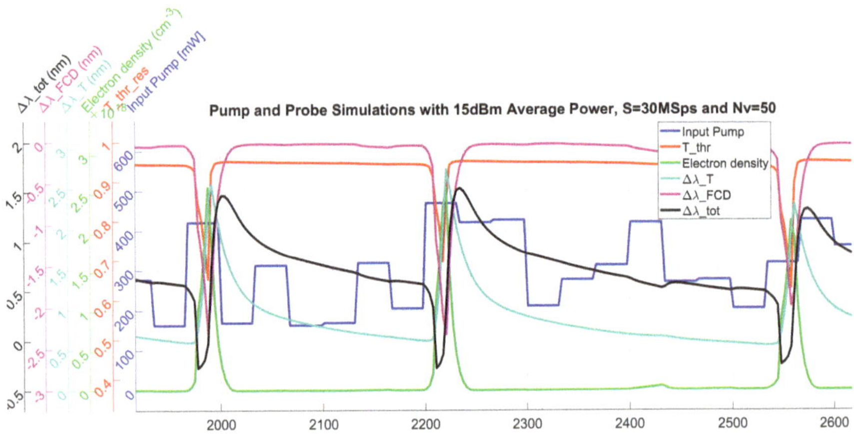

Fig. 26.5 Time evolution of the input pump power (blue line), the transmission coefficient T_{thr} at resonance (red line), the electron density (green line), the resonance wavelength shift due to temperature variation $\Delta\lambda_T$ (magenta line), the shift due to FCD $\Delta\lambda_{\text{FCD}}$ (cyan line), and the total wavelength shift $\Delta\lambda_{\text{tot}}$ (black line)

is nearly zero, leading to a maximum transmission coefficient at the cold probe resonance wavelength. In this state, the pump is no longer able to efficiently transfer information to the probe. However when the resonant wavelength returns close to the cold wavelength the circulating power increases and the cycle starts again.

26.4 Conclusions

In conclusion, we demonstrate the analysis on how non-linearities in silicon microrings can be exploited for reservoir computing. Our model, supported by experiments, is helpful in predicting the performance of the ring pumped by a pump signal that contains the data trace. Since the recombination of free carriers via silicon traps determines the carrier lifetime and hence the processing speed, we will explore also other ring designs with shorter carrier lifetime but yet high enough non-linearities.

This work is supported by Italian National Recovery and Resilience Plan (NRRP) of NextGenerationEU: partnership on Telecommunications of the Future (PE00000001—program "RESTART").

References

1. Borghi, M., Biasi, S., Pavesi, L.: Reservoir computing based on a silicon microring and time multiplexing for binary and analog operations. Sci. Rep. **11**, 15642 (2021). https://doi.org/10.1038/s41598-021-94952-5
2. Novarese, M., Romero Garcia, S., Cucco, S., Adams, D., Bovington, J., Gioanini, M.: Study of nonlinear effects and self-heating in a silicon microring resonator including a Shockley-Read-Hall model for carrier recombination. Opt. Express **30**(9), 14341–14357 (2022)

Chapter 27
Design of an Integrated Microlens Array for a High-Speed Photodetector Array

Yunjie Yan, Chengpeng Kang, Zhenlin Wu, Mingshan Zhao, and Geert Morthier

Abstract Free space optical (FSO) communication systems require high-performance receivers capable of converting high-frequency signals with wide bandwidths to support increasing data rates. An integrated photodetector array (PDA) enhances data reception efficiency by reducing receiver size and increasing the number of array elements. However, the small effective receiving area of single mode optical fibers poses significant alignment challenges and needs precise beam coupling. To address this, a microlens array (MLA) assisted PDA is proposed to improve alignment tolerance, reduce coupling losses and enhance optical coupling efficiency. This study explores the principles and design of a microlens array for a 10-element PDA operating at 1550 nm. The microlenses are arranged in a hexagonal configuration, which effectively manages incident light and compensates for alignment deviations. A spherical microlens with a radius of curvature of 2.4 mm is designed to optimize beam collimation and redirection, reducing lateral displacement. The results demonstrate that the MLA assisted PDA significantly enhances optical coupling efficiency, improving overall system performance and reliability in FSO applications.

27.1 Introduction

In free space optical (FSO) communication systems, rising internet data traffic has placed greater demands on data centers, compelling them to enhance their infrastructure and optimize their systems to handle an ever-growing digital load [1–3]. Therefore, the receiver should be capable of receiving high-frequency signals with a wide bandwidth to support higher data rates. An integrated photodetector array

Y. Yan · C. Kang · Z. Wu · M. Zhao
School of Optoelectronic Engineering and Instrumentation Science, Dalian University of Technology, Dalian, China

Y. Yan · G. Morthier (✉)
Photonics Research Group, INTEC Department, Ghent University, Ghent, Belgium
e-mail: Geert.Morthier@ugent.be

P. Smowton and T. Krauss (eds.), *Proceedings of the 26th European Conference on Integrated Optics*, Springer Proceedings in Physics 440,
https://doi.org/10.1007/978-3-032-29673-3_27

(PDA) helps to reduce the receiver size and increase the number of array elements, thereby enabling more efficient transmission of large scale data traffic. A wide field of view (FoV) is required in the receiving scenario [4]. The requirement improves signal coverage and it also enhances alignment tolerance. The system maintains a stable connection even when the incident beam shifts, thereby improving overall system reliability. However, when the receiver uses an optical fiber, especially a 10 μm single mode fiber, the effective receiving area is very small, requiring extremely high alignment accuracy to precisely couple the incident beam into the fiber [5]. An alignment tolerant interface is key for reducing the costs of photonic packaging. Besides, active alignment significantly increases both cost and assembly time during overall integration. Increasing the beam field diameter when spatial light is incident on the receiver, or enlarging the photosensitive area of the photodetector, can reduce alignment difficulties and coupling losses. Therefore, a microlens array assisted PDA can be employed for signal reception to meet the high precision and high efficiency application requirements of FSO systems.

In this work, we explore the use of a microlens array as a tool to address challenges in PDA. We introduce the principles of microlenses and provide a detailed design for their realization. We also explain how a microlens array manages light incident at various angles onto the photodetector array. By carefully designing the microlens array, a 10-element device was developed for a PDA at 1550 nm.

27.2 Principle

In this work, the microlens array is composed of 10 microlenses and the PDA is operating in back incidence at 1550 nm. To pack more tightly, the microlens array (MLA) elements are arranged in a hexagonal configuration. While the PDs are on the front side of the array, microlenses are integrated on the same side of it. For a simple deriviation of the projection formula, only the chief ray is considered as this passes through the center of the optical stop. As shown in the figure, a plano-convex microlens is characterized by its lens radius (r), vertex height (h), spherical crown curvature radius (R), refractive index (n), contact angle (θ) and focal length (f). The radius of curvature and the arch height of microlenses are calculated by the relationship between its diameter, focal length and radius of curvature. The curvature radius R of the microlens spherical crown can be expressed as [6]:

$$R = \frac{r^2 + h^2}{2h}$$

The focal length f of the microlens can be expressed as.

$$f = \frac{R}{n - 1}$$

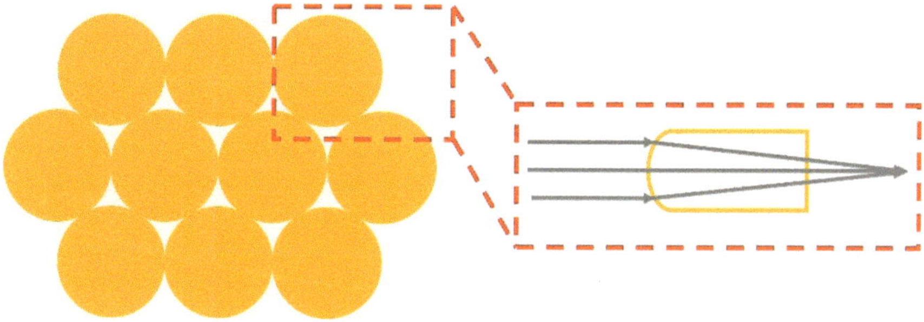

Fig. 27.1 Microlens array structure and single microlens model

n is the refractive index of the microlens material SiO_2 at a wavelength of 1550 nm. A schematic diagram of a 10-elements microlens array (a hexagonal (right) packed array of circular microlenses) is shown in Fig. 27.1. The fill factor η for the array is given by

$$\eta = \frac{\pi r^2}{p^2}$$

where p is the microlens pitch and r is the radius. The gap between the microlenses is given by$p - 2r$.

The microlenses array is packaged with the PDA, where the PDA adopts a back-illuminated configuration. The focal length of the microlenses must match the substrate of the photodetector to ensure that the converging light passing through the microlenses can be incident on the photosensitive surface of the PDA. Therefore, it is necessary to design a suitable microlens array to improve the optical coupling efficiency of the device and compensate for alignment deviations.

27.3 Results and Discussion

The PDA has an element pitch of 150 μm, a photosensitive area diameter of 50 μm and a substrate-to-surface distance of approximately 120 μm. Considering practical factors such as packaging and microlens fabrication processes, the focal length of the microlens needs to be greater than 3 mm. Therefore, from Fig. 27.2, the parameters with a ROC of 2.4 mm and a focal length of 5.4 mm were selected for further optical simulation.

Since the received light is normally a wide free-space beam, the light from the optical fiber is expanded to obtain a beam waist radius of 50 μm. The simulation is performed using Ansys Zemax OpticStudio [7]. In this structure, the beam is collimated and redirected by a single microlens, reducing the lateral displacement

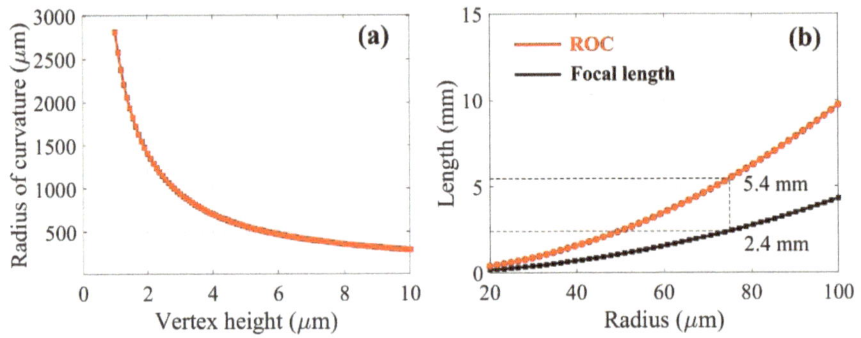

Fig. 27.2 Simulation results of **a** relationship between vertex height and ROC. **b** microlens radius with ROC and focal length

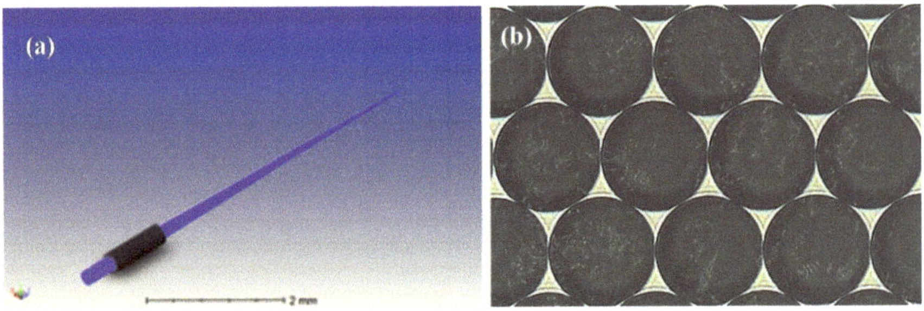

Fig. 27.3 Single microlens ssimulation model and fabricated microlens array

relative to the beam propagation axis. Figure 27.3 shows the simulation model and fabricated microlens array based on the designed parameters.

When the incident beam size is 100 μm (as shown in Fig. 27.4a), it is reduced to 45 μm after passing through the microlens, effectively converging within the photosensitive area (50 μm) of the photodetector. When the beam is incident at an angle of 0°,3°, and 5°, the beam size remains unchanged and still focuses within the photosensitive area. Figure 27.4 b, c, and d show that under different incident angles, the beam can still converge onto the photosensitive surface.

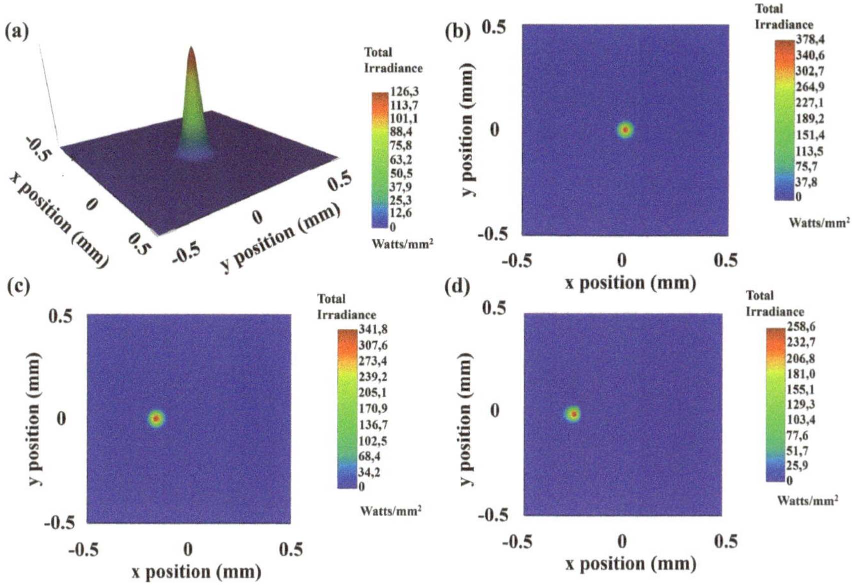

Fig. 27.4 **a** Gaussian beam under 0°incidence. Convergence of Gaussian beams at different incident angles after passing through the microlens **b** 0° **c** 3° **d** 5°

27.4 Conclusion

In this study, we designed and fabricated a microlens array for a 10-element photodetector array, effectively reducing the impact of incident light deviation on device performance. Simulation results show that under normal incidence and oblique incidence of 3°–5°, the microlens achieves a converging function at a focal length of 5.4 mm, meeting the application requirements for assisting the photodetector in receiving spatial light.

Acknowledgments We gratefully acknowledge the financial support provided by the program of China Scholarship Council (Grant No. 202306060161) and the BOF fund of Ghent University.

References

1. Huawei: 6G The Next Horizon. https://www.huawei.com/en/huaweitech/future-technologies/6g-the-next-horizon. Last accessed 07 Mar 2025
2. Gui, G., Liu, M., Tang, F., Kato, N., Adachi, F.: 6G: opening new horizons for integration of comfort, security, and intelligence. IEEE Wirel. Commun. **27**, 126–132 (2020)
3. Lu, H.-H., Huang, X.-H., Tsai, W.-S., et al.: 800 Gb/s/200 m FSO link with 284 AWDM-PAM4 scheme and SLM-based beam tracking technology. Opt. Lett. **46**, 1269–1272 (2021)

4. Umezawa, T., et al.: Radio over FSO communication using high optical alignment robustness 2D-PDA and its optical path switching performance. J. Lightwave Technol. **39**(16), 5270–5277 (2021)
5. Mangal, N., et al.: Monolithic integration of microlenses on the backside of a silicon photonics chip for expanded beam coupling. Opt. Express. **29**(5), 7601–7615 (2021)
6. Nussbaum, P., Voelkel, R., Herzig, H.P., et al.: Design, fabrication and testing of micro-lens arrays for sensors and microsystems. Pure Appl. Opt. **6**(6), 617 (1997)
7. Ansys Ansys Zemax OpticStudio. https://www.ansys.com/products/optics/ansys-zemax-opticstudio. Last accessed 07 Mar 2025

FSC
www.fsc.org
MIX
Papier | Fördert gute Waldnutzung
FSC® C083411

Kolibri 360 GmbH
Ferdinand-Jühlke-Straße 7
99095 Erfurt, Deutschland
produktsicherheit@kolibri360.de